PYCNOGONIDS

PYCNOGONIDS

P. E. King
Senior Lecturer in Zoology,
University College of Swansea

HUTCHINSON OF LONDON

HUTCHINSON & CO (*Publishers*) LTD
3 Fitzroy Square, London W1

London Melbourne Sydney Auckland
Wellington Johannesburg Cape Town
and agencies throughout the world

First published 1973

© P. E. King 1973

This book has been set in Times type, printed in Great Britain
on smooth wove paper by Anchor Press, and
bound by Wm. Brendon, both of Tiptree, Essex

ISBN 0 09 116460 5 (cased)

CONTENTS

1	Introduction	7
2	External morphology	9
3	Internal anatomy	41
4	Life cycle	72
5	Feeding	84
6	Geographical distribution	99
7	Affinities and evolution	112
8	Systematics	115
	Order Pantopoda	116
	Family Colossendeidae	
	Family Pycnogonidae	
	Family Endeidae	
	Family Pallenidae	
	Family Nymphonidae	
	Family Phoxichilidiidae	
	Family Ammotheidae and Tanystylidae	
	Order Palaeopantopoda	128
	Extra-legged species of the families Pycnogonidae, Nymphonidae and Colossendeidae	
	Key to the adults of pycnogonid families	132
	Bibliography	133
	Index of subjects	139
	Index of species	141

I

INTRODUCTION

The Pycnogonida are a group of marine arthropods often referred to as sea spiders because of their superficial resemblance to the true spiders or Araneae. Although various affinities have been ascribed to them, they are now generally considered to have class status and since there is almost no fossil record, any proposed relationships with other groups must be speculative. Earlier work on the pycnogonids was reviewed by Thompson (1909), Helfer and Schlottke (1935) and Fage (1949) but no recent review has been written. The group was given the name Pycnogonida by Latreille in 1810 and subsequently called Podosomata by Leach (1815) and Pantopoda, because of the length of their legs, by Gerstaecker in 1863. Pycnogonida is now given precedence.

The pycnogonid body is considerably reduced and what is left is almost wholly derived from a fused head and thorax known as the prosoma. The opisthosoma or abdomen is represented only by a small unsegmented protuberance with the anus at its tip. The mouth is situated at the end of a proboscis which is not usually present in arthropods. The proboscis accounts for half the length of the body in some species. Associated with the prosoma are a number of legs, usually eight but in some species ten or even twelve, and in others varying combinations of a pair of palps, chelifores and ovigerous legs. There is doubt as to whether the forms with more than four pairs of legs and extra body segments represent polymorphic species which have eight legs or whether they are separate species. The presence of these extra segments and the ability to regenerate limbs for a considerable part of the developmental period, though not as an adult, pose a number of interesting problems concerning the developmental process.

Problems of species determination are created by the reduced state of the body which creates a paucity of morphological characters which can be used by the taxonomist.

The pycnogonids have a wide geographical and bathymetric range, being represented in both polar and tropical seas and in both the inter-tidal zone and at depths greater than 6000 metres. They are usually only present in small numbers but in some places, where their main food supply—hydroids or polyzoans—is abundant, they occur in much larger numbers. Their effect as predators on the ecosystem has not been determined, but at depths where few other animals occur, or in regions such as the Antarctic where they are abundant, their role must be significant.

Most species of pycnogonid are holobenthic, living on the sea-bed throughout their lives and although a few species can swim, most can only crawl slowly over the substratum. These extremely limited powers of locomotion effect very little dispersion, thus increasing the chances of group isolation and possible speciation. The immature stages are not free living and are carried on the ovigers (Fig. 1) of the male so that they are also restricted by the limited movement of the adults. In some groups speciation has occurred, but the resultant forms only differ slightly from each other and the determination of species is very difficult without breeding experiments which have not yet been done.

The internal anatomy is simple, since to date there has been no excretory or respiratory system described and others, such as the circulatory system, are simple. The pycnogonids are unique in their method of intracellular digestion. Certain of the mucosal cells take up food material until they are gorged with food and they then strip away from the mucosa and float freely in the lumen of the midgut. Other mucosal cells absorb food from them until, with the food exhausted, the floating cells are eliminated from the anus. Many other questions regarding pycnogonid physiology, and indeed all aspects of their biology, are unanswered and these interesting animals deserve rather more attention than has been lavished on them to date.

2

EXTERNAL MORPHOLOGY

The body of a pycnogonid consists of a proboscis, a cephalothorax with a number of appendages and an abdomen, or anal appendage, which is invariably greatly reduced and in living species consists of a single segment (Fig. 1). The anterior part of the cephalothorax which joins the proboscis constitutes the head or cephalon and in most species has a tubercle on its dorsal surface bearing the eyes. The cephalon and the first trunk segment are fused in all species but in some, e.g. *Nymphon gracile,* the other trunk segments are externally distinct. The cephalothorax bears a number of appendages and in some species, e.g. *N. gracile,* there is the full complement of these consisting of a pair of palps, chelifores, ovigerous legs and ambulatory or natatory legs. The first two, the palps and chelifores, are not present in all species and the ovigerous legs are frequently only present in the male, though in some species, e.g. *N. gracile, Achelia echinata,* both sexes have them, those of the female being smaller than those of the male.

Each segment of the thorax has a prominent lateral process with which the legs articulate. Most species have four pairs of legs though some species, e.g. *Pentapycnon charcoti,* have five, and a few have six pairs.

The whole body is covered with cuticle which, on its outer surface, is either smooth, tuberculated, with scattered setae or sometimes covered with a dense coat of setae.

In many species, the segmentation of the thorax is externally visible but in others, e.g. *Achelia laevis,* some or all of the sutures are indistinct externally. In some the adjoining lateral processes are clearly separated from each other but in others, such as *Clotenia conirostris,* in which the body is disc-like, they are close together and in some instances may be fused.

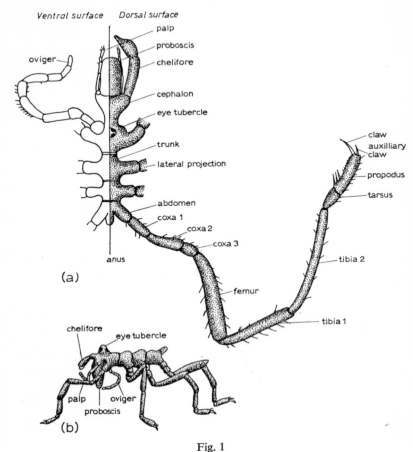

Fig. 1
(a) Diagrammatic representation of a pycnogonid showing the parts of a typical specimen from the dorsal and ventral aspects. Only one walking leg is shown and only one of each pair of the other appendages. This body form is typical of the Nymphonidae.
(b) Diagrammatic representation of the lateral aspect of a typical member of the Ammotheidae.

External morphology

Most pycnogonids move by slowly crawling over the substratum though the extent to which this occurs has never been determined. Some of the species with an elongate body and limbs can swim, e.g. *N. gracile*.

THE PROBOSCIS

The most prominent external feature of all pycnogonids is the proboscis which varies considerably in shape and size between species (Fig. 2). This structure bears the mouth at its distal end and thus is important as the principal food-gathering organ. Its shape and mode of action are correlated with the method of feeding in individual species. The proximal end of the proboscis is attached to the cephalothorax by a ring of flexible cuticle usually referred to as arthrodial membrane or the 'soft collar'. The degree of movement of which the proboscis is capable in relation to the rest of the body is partly determined by the width of this collar. In species which are capable of considerable movement the collar is usually relatively wide and in those which are not capable of much movement the collar is narrow. The proboscis is moved in relation to the rest of the body by groups of muscles collectively referred to as extrinsic muscles. The proboscis gathers food by the rasping action of the jaws, situated around the mouth coupled with a sucking mechanism implemented by distortion of the pharynx. The muscles responsible for this action attach to the wall of the pharynx and the cuticle of the proboscis, and are collectively referred to as the intrinsic muscles. In pycnogonids (e.g. *Pycnogonum littorale, Pycnogonum planum* and *Austrodecus breviceps*), which attach to relatively large animals such as actinians or sponges, or have well developed chelifores which gather the food and carry it to the mouth as in *Nymphon gracile, Callipallene brevirostris* or *Parapallene capillata,* the proboscis is usually incapable of large movements and can be held rigidly along the axis of the body or at some definite angle to it by the extrinsic muscles. In other species with different sources of food, such as detritus-gathering by *Endeis spinosa,* grazing on algae by *Achelia longipes* or attacking prey with the proboscis such as *Achelia echinata* on *Flustra foliacea,* all of which lack chelifores, the proboscis is capable of considerable movement at the region of the 'soft collar', and in some examples, *Ascorhynchus auchenicum* and *Achelia echinata,* may be directed downwards at right angles to the main axis of the body or even in a more ventro-posterior direction than this, a characteristic of many Ammotheids (Fig. 2). The surface of the cuticle covering the proboscis may be smooth, partly or wholly covered with tubercles as in *Pycnogonum littorale* or with one or two projections as in *Pycnogonum rhinoceros.* There are usually a number of bifid setae

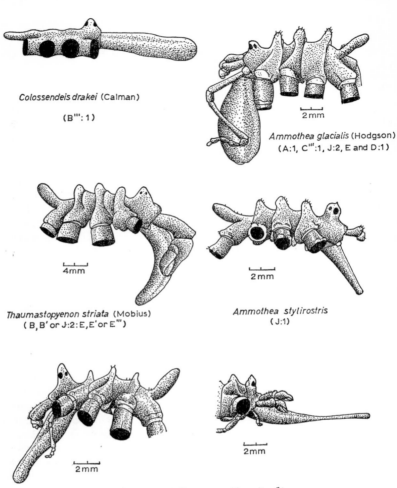

Fig. 2

A series of lateral aspects of pycnogonids with the ambulatory appendages removed to illustrate the range of shape and orientation which the proboscis may assume and its position relative to the other appendages of the cephalon. The letters and figures denote the classification of the shape of the proboscis as suggested by Fry and Hedgpeth (1969), see Fig. 3. (Redrawn from Fry and Hedgpeth, 1969)

scattered on the surface and sometimes groups of tactile or chemosensory spines. These are particularly abundant around the mouths of detritus-feeding forms such as *Endeis spinosa* and *Endeis laevis*. In some species the shape or relative size of the proboscis of the juvenile differs from that of the adult. In *Ammothea longispina* the proboscis of the juvenile is more elongate than that of the adult (Fig. 2) but in most examples in which the proboscis of the juvenile differs from that of the adult the juvenile one is shorter and broader. This is characteristic of many members of the Nymphonidae such as *Nymphon gracile* and *Nymphon rubrum* and has led to confusion regarding the validity of related smaller species such as *Nymphon brevirostre* which occur in a similar habitat and have sometimes been considered as juvenile forms of the others. Doubt still exists regarding species in the Russian Arctic but around British coasts these three species are distinct.

In some species it is possible that the food source of the juvenile differs from that of the adult. In members of the families Endeidae and Ammotheidae the juveniles have chelate chelifores but in the adults of the ammotheids the chelae atrophy and in the Endeidae the whole chelifore is shed.

The descriptions of the shape of the proboscis using single adjectives such as 'clavate', 'oval' or 'spindle-shaped' can include a wide range of variations in form and thus they frequently obscure a number of characteristic shapes which are used in diagnosing the affinities between species or higher taxa.

Fry and Hedgpeth (1969) proposed a scheme of proboscis classification which utilised a number of characters (Fig. 3). These include the proximal and distal diameters of the whole proboscis; the presence or absence of a dilation at some point along the proboscis and, if present, the position of such a dilation with its relationship to the mid-length point of the proboscis; the presence or absence, and the size relative to the proximal diameter, of a second and more distal dilation; and the type of curvature of the proboscis. The assessments of curvature are made from the lateral aspect (Fig. 3b), and the other criteria are determined from the dorsal aspect (Fig. 3a). In Fig. 2 a number of different proboscides are shown together with their description in terms of these characters.

Structure of the proboscis

A detailed study of the structure and function of the proboscis has been made only on some British and Antarctic species. In the Antarctic species *Pycnogonum stearnsi* and *Rhynchothorax australis* the wall of the proboscis is composed of uniformly non-flexible cuticle, but in *Austrodecus glaciale,* with an attenuated distal portion, this part of the proboscis has a wall of the same thickness overall but it is composed of flexible cuticle. Interspersed in the flexible cuticle of this proboscis, there are rings of non-flexible cuticle to which muscles are attached.

The British species all have proboscides covered with cuticles of uniform thickness over the whole proboscis. Several species of the genus *Pycnogonum*—*P. hancocki* from the Galapagos Islands and *P. rickettsi* from California—have a reticulation of thickened ridges internally. These are not confined to the proboscis but occur on other parts of the body. They do not have muscles attached and probably give added rigidity to the cuticle. These species have relatively short powerful legs and feed by pushing the tip of the proboscis against the host tissues. In a related species which lacks the thickening, *P. littorale,* the tip of the proboscis is small and bulb-like so that once it is inserted in the host tissues they close round it and help to anchor the mouth in position to maintain suction. In the forms with the reticulations the proboscis has a similar diameter all down its length so that pressure must be exerted continually.

The mouth at the tip of the proboscis is a triradiate structure which usually has three lip-lobes. These are developed to varying degrees in different species, and sometimes consist of smooth cuticle as in *Achelia echinata,* a number of small tubercles, *Pycnogonum littorale,* or carry a number of setae as in *Endeis spinosa*. In most species, the three lips are of approximately the same size but in *R. australis* the dorsal lip is much smaller than the ventrolateral lips. In most species, the interradial portions of the end of the proboscis are composed of arthrodial membrane while the radial ridges of the foregut are firmly anchored to the outer proboscis wall by hardened cuticle. Near the mouth, the interradial grooves of the pharynx are

---

Fig. 3

Diagrams showing criteria used by Fry and Hedgpeth (1969) to describe the shape of the pycnogonid proboscis. (Redrawn from Fry and Hedgpeth, 1969)

(a) The five main proboscis shapes in the Pycnogonida, A, B, C, D and J. The variations on the basic them are denoted as A, A'; B, B', B'', etc. The criteria involved are: the presence or absence and position of one or two dilations; the relative diameters of the proboscis and its insertion into the cephalic somite X, and at its distal extremity Y. Z indicates the midpoint of the longitudinal axis.

(b) Types of proboscis curvature from the lateral aspect. Types of variation in the distribution of the curvature:

    E = uniform throughout
    E' = greatest curvature about midpoint of the axis
    E'' = greatest curvature proximal to the midpoint
    E'''= greatest curvature distal to the midpoint
    Z = midpoint of axis length

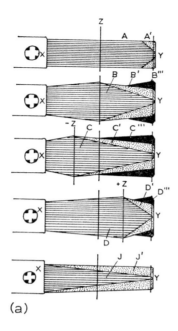
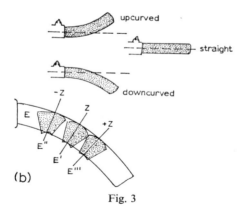

Fig. 3

thickened and produced to form apophyses. To these are attached the tendons of the lip muscles. *R. australis* has no apophysis on the dorsal lip. In some species lip movement is an important part of food gathering, e.g. *A. echinata* thrusts the tip of its proboscis through the operculum of the bryozoan *Flustra foliacea* and feeds on the animal within (Fig. 27d,e,f).

The part of the alimentary canal within the proboscis, which is usually called the pharynx, is lined with cuticle, which is continuous anteriorly with the outer cuticle of the proboscis (Fig. 4a,b). It is lost and re-formed at ecdysis. Throughout its length the foregut is trifoliate in cross-section and in regions of greatest cross-sectional area the shape expands and may approach that of an equilateral triangle. This reflects the basic triradial symmetry of the proboscis. The ridges of cuticle which mark the boundaries between the cuticle of each antimere are usually referred to as radials and the part covering an antimere is considered to be the interradial. The radial ridges, in all species examined, are thicker than the interradial walls, which each have a longitudinal groove with a thickened floor (Fry, 1965). The proximal part of the pharynx wall has a number of thickenings forming annular bands. These bands have long, fine setae which project forwards and slightly inwards (Fig. 4a). This structure is believed to be a mechanism for macerating tissue ingested by the pycnogonids, and in this role it may be assisted by more anteriorly situated stout teeth, e.g. in *Endeis spinosas* (Dohrn, 1881) (Fig. 4b,d). It may also act as a filter to prevent the passage of large, and possibly hard, objects into the narrowed posterior end of the pharynx and thus into the midgut into which it empties through a narrow tube. The mechanism of the proboscis depends upon the interaction of the hard parts and the musculature.

Dohrn (1881) and Hoek (1881) gave descriptions of the musculature of the proboscis but did not suggest details of the functioning of individual muscles or groups of muscles. Fry (1965) gave a detailed description of these muscles in three Antarctic

---

Fig. 4

(a) Diagrammatic representation of the longitudinal section of the body showing the position of the alimentary canal in relation to the other internals and inset a longitudinal section of the proboscis showing the position of the setae and the oesophagus.
(b) Longitudinal section through one section of the proboscis of *Endeis spinosus* (after Dohrn 1881).
(c) Lip and jaw associated with the mouth region of one segment.
(d) Arrangement of setae on the supporting bars inside the pharynx.
(e) Dorsal aspect and longitudinal section of the proboscis of *Pycnogonum hancocki* showing the reticulations formed by thickenings of the cuticle.

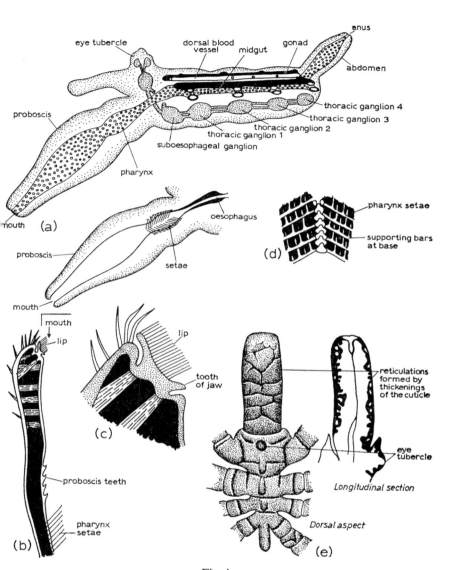

Fig. 4

species, *Rhynchothorax australis, Austrodecus glaciale* and *Pycnogonum stearnsi,* and Wyer and King (in prep.) in a number of British species. All the species differed in the details of muscle insertion but the basic pattern was similar.

Intrinsic musculature of the proboscis

In pycnogonids so far examined, there are four groups of functionally distinct intrinsic muscles. These are the muscles associated with the mechanism of sucking and mastication.

(1) The radial muscles, usually present along the entire length of the proboscis, connect the thickened radial ridges of the foregut to the inner surface of the proboscis wall (Fig. 5).

(2) The interradial muscles are attached on either side of the thickened interradial grooves of the foregut and pass outwards to the inner surface of the foregut wall. In some species these muscles have been depicted as overlapping in their regions of insertion on the foregut wall (Fig. 5).

(3) The circular muscles are present only in the proximal part of the foregut where its cross-sectional area decreases abruptly. Each band of muscles consists of three separate arcs of fibres, each of which are attached at both ends to the thickened radial ridges of the foregut. These rings of circular muscles alternate with rings of radial and interradial muscle fibres. Some early researchers interpreted these circular muscles as nerve fibres, but they have since been described as striated structures which indicates that they are muscles (Fry, 1965) (Fig. 5).

(4) The lip muscles originate on the inner surface of the proboscis wall and run obliquely forwards to their insertions on a tendon which is attached to the large lip apophyses of the two ventrolateral units. In a few species a similar muscle, tendon and lip apophysis lie in the

---

Fig. 5

Diagrammatic representation of the transverse sections of the proboscis.
(a) An interpretation of the proboscis structure of *Nymphon robustum* by Hoek (1881).
(b) An interpretation of the proboscis structure of *Trygaeus communis* and *Endeis spinosus* by Dohrn (1881).
(c) Section through the proboscis at the level of insertion of the proboscis into the cephalic somite of *Rhynchothorax australis*.
(d) Section through the proboscis of *Rhynchothorax australis* more distal to (c).
(e) Section through the proboscis of *Rhynchothorax australis* at the level of the main proboscis ganglion.
(f) Section of proboscis of *Rhynchothorax australis* immediately behind the mouth. (Redrawn from Fry, 1965)

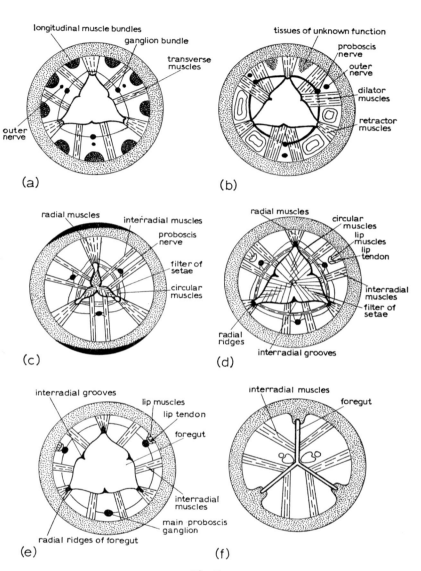

Fig. 5

dorsal longitudinal segment. The muscles and their tendons lie between the interradial muscles of each longitudinal segment. In some species, researchers have described modified interradial muscles as lip muscles, e.g. *Austrodecus glaciale,* but more information is needed on this point (Fig. 5) (Fry, 1965).

Functioning of the proboscis
The basic action of the proboscis—the sucking and masticatory organ of pycnogonids—depends upon the antagonism of the elasticity of the foregut wall and the radial and interradial muscles.

Contraction of the interradial muscles pulls the interradial walls of the foregut towards the wall of the proboscis. Simultaneous contraction of the radial muscles creates tension in the radial and interradial ridges, and the action of the radial muscles keeps the whole foregut in position within the lumen of the proboscis. The result of the contraction of these two sets of muscles is that the cross-sectional shape of the foregut changes from trifoliate to more nearly triangular and, as this shape changes, so the volume of the foregut increases. This in turn produces a lowering of the pressure in the lumen of the foregut resulting in an inrush of material from outside the animal when the lips are open. Relaxation of the radial and interradial muscles releases the tension on the grooves and ridges of the foregut which tends to allow the foregut to resume its trifoliate cross-sectional shape. This reduces the volume of the foregut lumen so that pressure builds up on the ingested material. Then, provided that the lips of the mouth are securely closed, the foregut contents will be forced backwards.

The action of the muscles on the spines of the sieve apparatus is such that when the pharynx is in the dilated condition it is completely blocked to large particles, the spines being directed towards the mouth and thus preventing the penetration of food into the oesophagus (Fig. 4a). When the pharynx is contracted, the spines lie flat against the intima. The action of the spines against one another causes a mechanical breaking up of the food, even if it is in a fairly diffuse form. The resistance and shearing action of the spines causes the food to be put under pressure and so broken up before passing into the midgut (Fry, 1965).

Extrinsic musculature of the proboscis
Whilst the arrangement of the intrinsic musculature of the proboscis is clearly based on a triradial arrangement, the muscles responsible for movement of the entire proboscis, termed the extrinsic muscles, are not arranged in this pattern. They consist of at least two pairs of well developed muscles, which in some species are subdivided

(Wyer, 1972). Occasionally there is an additional single smaller pair of muscles in a ventral position (Fig. 6). In *R. australis* three pairs of muscles are involved. The first pair of muscles is responsible for retraction of the dorsal proximal end of the proboscis and is attached to a transverse dorsal apophysis, medial to the origins of the second pair of muscles. They then pass anteriorly through the circumoesophageal commissures between the dorsal ganglion and the dorsal surface of the gut (Fig. 6) and attach to the anterior dorsal surface of the arthrodial membrane which is situated between the proboscis and the cephalic somite. As they pass forwards the muscles converge on each other so that at their anterior insertion the right and left components of the pair are nearer together than at the points of their posterior insertion. The second pair of muscles has its posterior origins on the same transverse apophysis as the first pair but their points of attachment are lateral to those of the first pair of muscles. The second pair of muscles passes forwards and downwards passing outside the circumoesophageal commissure, and is inserted on the ventrolateral walls of the arthrodial membrane between the proboscis and the cephalic somite. Unlike the first pair of muscles, these diverge slightly as they pass forwards.

In *R. australis* the third pair of muscles is present. They are considerably shorter and of smaller diameter than the other muscles involved (Fig. 6). Their origins lie on the proximal ventral portion of the arthrodial membrane between the proboscis and cephalic somite and have their anterior insertions on a low apophysis on the ventral proximal edge of the proboscis wall. In this species and the other two studied, there are no visible muscle antagonists to these sets of muscles and it is assumed that the body fluids are the antagonists of all the extrinsic proboscis muscles (Fry, 1965).

Fry suggests that during life the system is in a balanced state when all the extrinsic muscles have a slight tonus, and the proboscis is held with its longitudinal axis roughly parallel to the longitudinal axis of the body. When the first pair of muscles contracts, the proximal dorsal surface of the proboscis is pulled backwards and the dorsal portion of the arthrodial membrane will buckle inwards. The whole proboscis will then tilt upwards at the distal tip by a folding of the arthrodial membrance. When the second pair of muscles contracts, the proximal ventral edge of the proboscis is pulled backwards and upwards resulting in the lowering of the distal end of the proboscis. The angles which the pairs of muscles make with the longitudinal axes of the body and proboscis suggest that the second pair of muscles produces much greater movement from a balanced state than the first pair, and by their divergence are capable of producing some lateral movement of the base of the proboscis if

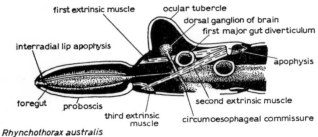

Rhynchothorax australis

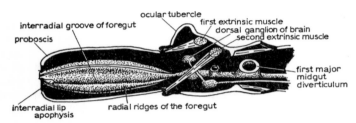

Pycnogonum stearnsi

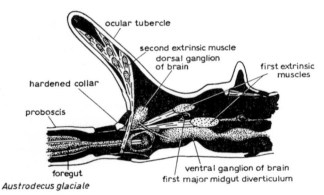

Austrodecus glaciale

Fig. 6

Diagrams showing the arrangement of the extrinsic muscles in *Rhynchothorax australis*, *Pycnogonum stearnsi* and *Austrodecus glaciale*. (Redrawn from Fry, 1965)

the components either contract independently of each other or to a different extent. The first pair of muscles has several functions. They aid the hydrostatic pressure in returning the system to a balanced state when the second pair of muscles relaxes after a major contraction. When the first and second pairs of muscles contract slightly the balance of the system will be maintained and when they are strongly contracted they also act to slightly retract the proboscis.

Pycnogonum stearnsi has only two pairs of muscles involved in moving the proboscis. The first pair of muscles, however, does not pass between the circumoesophageal commissures (Fig. 6). The first pair of muscles has its origin on the dorsal wall of the cephalic somite, partly inside the ocular tubercle. The muscles pass downwards in an anterior direction and are attached to the distal dorsal edge of the arthrodial membrane between the proboscis and cephalic somite.

The second pair of muscles is attached posteriorly to the lateral walls of the cephalic somite, posterior to the regions of origin of the first pair. They then pass anteriorly in a downwards direction, outside the circumoesophageal commissures and insert on the ventrolateral walls of the arthrodial membrane.

The raising and lowering of the tip of the proboscis is apparently brought about by the contraction of the first or second pairs of muscles respectively, but the relatively small angles which the muscles make with the longitudinal axis of the body suggest that vertical movements of the proboscis are unlikely to be very powerful. Contraction of both pairs of muscles together tends to retract the whole proboscis and it is reasonable to suppose that the proboscis is withdrawn periodically during feeding since this species, in common with most others of the genus *Pycnogonum,* feeds on large, soft, food material such as anemones.

Austrodecus glaciale also has only two pairs of muscles involved in moving the proboscis, but the region of articulation between the proboscis and the cephalic somite is more complicated than in either *R. australis* or *P. stearnsi* (Fig. 6). The unsclerotised cuticle joins the hardened part of the cephalic cuticle in a ventral position.

The first pair of muscles is anteriorly attached to the arthrodial membrane immediately behind the proximal dorsal edge of the proboscis. These pass backwards between the circumoesophageal commissures. The right and left components of the pair each divide into two, and these diverge slightly to attach on the lateral walls of the cephalic somite. The second pair of muscles is much larger than the first and is attached over almost the whole wall of the ocular tubercle which is considerably elongated in this species. They then pass almost straight downwards, outside the circumoesophageal commissure and insert on the ventrolateral walls of the arthrodial membrane, immediately posterior to the proximal edge of the proboscis.

CHELIFORES

The term 'chelifore' is used since 'chelicera' has connotations regarding relationship with the Arachnida and it is undesirable to imply an unproven relationship of this nature. The chelifores when present are inserted on each side of the cephalon. They are typically composed of a number of segments two of which form a basal unit referred to as the scape, for example as in the *Decolopoda* and some species of *Ascorhyncus*. In *Nymphon, Phoxichilidium, Callipallene* and *Cordylochele* (Fig. 7d) the scape is formed from one segment only. In those species with chelifores bearing chelae the two distal segments combine to form this structure. The 'movable finger' of the chela is formed from the ultimate segment of the chelifore and the immovable part by the penultimate segment. In some species of the genus *Nymphon*, the fingers of the chelae are greatly elongated and possess a formidable array of teeth on their opposing surfaces, whilst in others the fingers are much shorter and the teeth blunter (Fig. 7a). In *Anoplodactylus angulatus* there are a few relatively large teeth and in *Cordylochele longicollis* the hand is almost globular and the movable finger is very small and almost enclosed in a depression in the hand. In *Boreonymphon robustum*, the fingers are prominently curved, with a wide gap between them (Fig. 7b). The degree of development of this structure depends upon the method by which food is passed to the mouth. In the genera which have a well developed proboscis, e.g. *Ammothea, Colossendeis, Endeis* and *Pycnogonum*, the chelifores are absent or reduced since the proboscis is usually thrust into the tissues of the host or gathers food by its own action (Thompson, 1909).

The chelifores are absent in the adults of *Pycnogonum, Endeis, Rhynchothorax* and *Colossendeis*. In the Ammotheidae, they are reduced and the chelae only represented by small knobs in the adult. In *Trygaeus* and *Tanystylum* the whole chelifore is reduced to a small knob. In the Ammotheidae, they are fully chelate in the larval stages, and are probably used to attach the larvae to hosts such as hydroids. In the Endeidae fully chelate chelifores are present in the immature stages but the whole chelifore is cast off when some growth has taken place but before the adult size is reached. It is possible that a change of diet or better developed legs to aid holding the substratum renders them superfluous in the adult. In *Eurycide* and *Ascorhyncus* they are comparatively small with imperfect chelae.

In those genera possessing well developed chelifores these structures

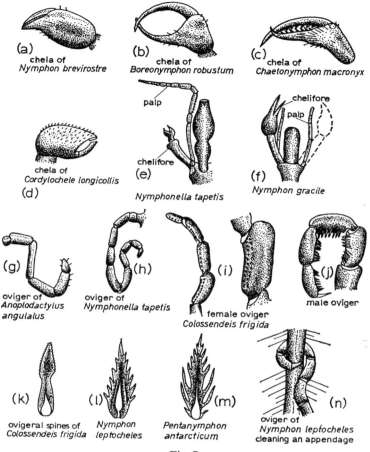

Fig. 7
(a)–(d) Examples of chelae with different shapes from chelifores.
(e)–(f) Palps. (e after Ohshima, 1927a)
(g)–(n) Ovigers. (h after Ohshima, 1927a; j after Bouvier, 1923; m and n after Prell, 1910)

are used either to grasp the prey before the proboscis is thrust into it or to seize the prey, probably hydroids, and thrust it into the mouth.

PALPS

The second pair of appendages when present is inserted at the side of and laterally to the chelifores (Fig. 1). The number of segments composing these appendages varies between species; ten segments are a large number, though a few species with this number exist. The palps of *Nymphonella tapetis* (Fig. 7e), a species recorded only from Japan and the Mediterranean, have twenty segments and those of *Eurycide* have seventeen, though these would appear to be exceptional conditions (Ohshima, 1927a). Palps are absent or vestigial in the adults of species belonging to the genera *Pycnogonum, Endeis, Phoxichilidium* and *Callipallene*. In *Phoxichilidium* the palps are represented by a knob and in *Pallenopsis* by a single segment. In some of the Pallenidae there is a sexual difference in that the reduction of the palp has gone further in the female than in the male. The significance of this is at present unknown. The relative lengths of the terminal segments are used to determine some of the species in the genus *Nymphon* (King and Crapp, 1971). The terminal segments of the palps are usually setose, the setae being presumably sensory and concerned with the tactile sense. In *Colossendeis,* the palp is bent in a zig-zag manner and similarly in *Achelia*. In the last named species the terminal segments of the palp are held alongside the tip of the proboscis and in *Achelia echinata* appear to assist in its orientation above the operculum of *Flustra foliacea*. In *Achelia echinata, A. longipes, A. hispida* and *A. simplex* the setae on the terminal segments have a small terminal pore and, although there is no direct evidence from these species, this type of structure is thought to have a chemosensory function in other anthropods. In *Nymphon gracile* (Fig. 7f) it is possible that the palps help the chelifores bring the food near to the proboscis in addition to their possible sensory function.

OVIGEROUS LEGS

Unlike the other appendages described above, these are attached to the ventral surface of the body, posteriorly situated in relation to the palps. They are absent in the females of the genera *Pycnogonum, Endeis* and *Phoxichilidium,* but in some species they are present in both sexes though frequently smaller in the female than the male (Fig. 8b). They first appear as a bud in the early larval stages and as development proceeds they increase in length, become segmented

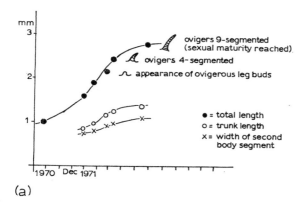

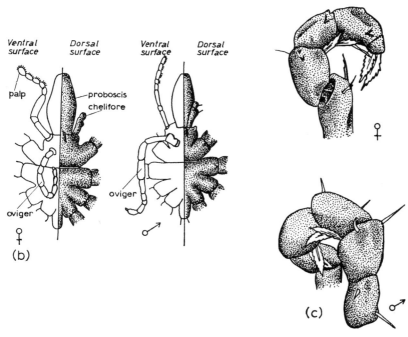

Fig. 8

(a) Size increase of male *Pycnogonum littorale* observed in an aquarium until sexually mature. ● = Total length; ○ = trunk length; × = width of 2nd body segment.
(b) Dorsal and ventral aspects of both sexes of *Achelia longipes* with the legs omitted showing the relatively larger size of the oviger in the male.
(c) Terminal segments of the male and female ovigers of *Achelia echinata* showing the differences in spines.

and eventually produce their full complement of articles (Fig. 8a). The number of these segments varies between species; in *Phoxichilidium* there are five, *Anoplodactylus* six, *Endeis* seven, *Pycnogonum* nine and *Colossendeis* ten. In some there are a number of serrated spines on the terminal segments and in many species, notably members of *Colossendeis* and *Achelia,* the terminal segments have an armature consisting of a variety of toothed denticles. These are usually set in a membranous socket (Fig. 7). When ovigerous legs occur in both sexes, those of the male usually have more of these serrated spines than those of the female (Fig. 8c). They are presumably used for cleaning whilst the short simple spines which predominate on the terminal segments of the male oviger are used to hold the balls of eggs. This is particularly evident in members of the genus *Achelia* (Fig. 8b,c).

When there is complete development of the ovigers, i.e. ten segments, the proximal three segments corresponding to the coxae of the ambulatory appendages are short. The next three segments are longer and the terminal four segments, which presumably represent a subdivision of the tarsus and propodus of the ambulatory appendages, are short. In species with long ovigers, these appendages have their terminal segments coiled spirally.

The appendages are used by the males to carry the eggs, after their release by the females, until the larvae hatch. An exception is *Colossendeis* in which no male has so far been found carrying eggs. The balls of eggs are held on the ovigers sometimes by a dilation on one of the distal segments of the oviger which corresponds to the first tibia of an ambulatory appendage or by the small spines. In some species the balls of eggs are aggregated by mucus or cement released by glands on the femur at the time of egg release. Bouvier

Fig. 9
(a) Leg of *Achelia echinata* with the usual number of segments present in the legs of pycnogonids.
(b) Front leg of *Nymphonella tapetis* with extra segments.
(c) Graph showing the changes in the dimensions of the femur of *Nymphon gracile* at different times of the year.
(d) The sequence of movements of the leg segments in relation to each other during the up and down movements of swimming. Phases 1–4 for the power stroke and 5–8 the recovery stroke.
(e) The position of the leg segments when held in the plummeting position. (Redrawn from Prell, 1910)
(f) Graph showing the relationship between the vertical swimming speed and the leg-beat frequency (●); the subzero points along the horizontal axis were recorded from animals walking on the bottom. Rates of sinking with legs extended (○) and in the plummeting position (△) are indicated at zero beats/sec. (Redrawn from Morgan, 1971)

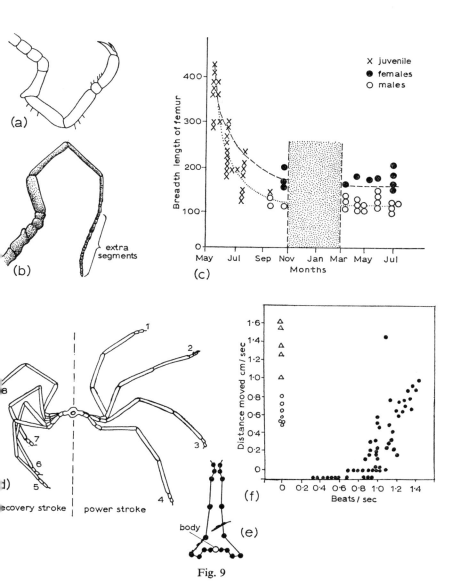

Fig. 9

considered that the egg-carrying function of the ovigers has been secondarily acquired and that the form in *Colossendeis* represents the primitive condition.

In some species, and in females which have them, the ovigers are used to clean the other appendages. They are wrapped around the bases of these appendages and drawn along the appendage to remove detritus or other accumulations on the surface of the appendages (Fig. 7l). When males of *Nymphon gracile* are carrying eggs they become covered with algae or encrusting polyzoans such as *Electra pilosa*. This condition presumably results from the ovigerous legs being unable to clean the body surface when they are carrying eggs. No investigation has been made of the possible presence of glands in some segments of either the palps or the ovigerous legs.

LEGS

The ambulatory or natatory appendages have nine segments. These are three proximal coxae, a femur, two tibiae, a tarsus, a propodus and a terminal claw (Figs. 1a,9a). The size of the legs varies from about equal in length to the body as in species of *Pycnogonum, Ammothea* and *Rhynchothorax,* to more than six or seven times body length as in species of *Nymphon, Endeis* and *Colossendeis* (Figs. 37,38,39,40). The femur and two tibiae are usually the ones which undergo greatest elongation in these forms though the tarsus, which in most species is short, is greatly elongated in some Nymphonidae. The propodus is usually somewhat curved and usually has an armature of simple or serrate spines. Near the base of the terminal claw in some species there is a pair of auxiliary claws in members of the Ammotheidae, *Phoxichilidium,* many species of the genus *Pallene* and nearly all Nymphonidae. The presence or absence and point of insertion of these auxiliary claws are frequently used as taxonomic characters.

An exception to the general pattern occurs in a Japanese and Mediterranean pycnogonid, *Nymphonella tapetis* (Ohshima, 1927a), the young of which live in the bivalves *Paphis philippinarum* and *Protothaca jedoensis.* In these, the terminal segments of the forelegs have subdivided to give a flagellum (Ohshima, 1927a) (Fig. 9b). The eggs are carried on the ovigers of the male and when they hatch, the male approaches the inhalant siphon of the *Paphia* and it is conjectured that the modified front legs and similarly modified palps are used to place the newly hatched larvae either into the stream of water or actually into the siphon and thus enable them to reach the mantle cavity of the shell.

In females of *Nymphon gracile* there is a seasonal variation in the

External morphology

dimension of the legs. The femora are proportionately much broader when mature eggs are present in them than when these are in an early stage of development (Fig. 9c). Some or all of the legs have the genital openings on their second coxae. The exact position and number depends upon the species involved (Table 1).

LOCOMOTION

Movement in pycnogonids may occur actively by walking or swimming or passively by being carried in ocean currents whilst attached to hydroids, weed or polyzoans. Although the usual active method of movement by pycnogonids consists of a slow crawl over the substratum the distance moved has rarely been determined and for many species movement is probably restricted to the clump of hydroids or polyzoa on which they were deposited as juveniles. Many of the more delicate genera with an attenuated body and long thin legs such as *Anoplodactylus, Pallene* (Cole, 1901) and *Nymphon* (Prell, 1910; Fage, 1932; Morgan, Nelson-Smith and Knight-Jones, 1964) have been reported swimming freely, propelling themselves through the water by means of the four pairs of walking legs. These beat mainly in a vertical plane in relation to the body axis so that the animal moves through the water dorsal side first (Morgan, 1971). In *Nymphon gracile,* the only species of pycnogonid in which swimming has been studied, the legs beat ventrally in a metachronal sequence starting from the rear (Knight-Jones and Macfadyen, 1959), but apart from being longer and more slender than the legs of the heavier built non-swimming species, they do not appear to be specialised for swimming as are for example the pereiopods of portunid crabs (Morgan, 1971). Neither the chelifores, palps nor ovigers make any contribution to swimming (Morgan, 1971).

With the exception of coxa 2, which articulates laterally, all the leg segments move in a vertical plane. The muscular organisation at the joint is basically similar to the crustacean pattern (Lochhead, 1961), with a pair of muscles arranged antagonistically in the plane of flexion at each joint. Coxa 1 moves both above and below the longitudinal axis of the joint and the levator and depressor muscles, which originate extrinsically in the lateral extension of the body segment, are equally developed. The adductor and abductor muscles of coxa 2, located laterally in the first coxa, are also equally proportioned. Coxa 3 and the femur articulate primarily in a dorsal arc during swimming and the fan-shaped levator muscles, which have their origins dorso-laterally in coxae 2 and 3 respectively, appear more powerful than their relatively slender antagonists.

In contrast it is the flexor (depressor) muscle of tibia 1 which is the

Table 1
Summary table

	Proboscis	Chelifores	Palps	Ovig. legs	Trunk segments	Genital openings Male	Female
Nymphonidae	Large fixed	Large scape 1—jointed	5	8–10 ♂, ♀	Well segmented	2,3,4,(5)	1,2,3,4,(5)
Colossendeidae	Somewhat mobile	In juvenile only	10	10 ♂, ♀	Coalescent	1,2,3,4	1,2,3,4
Ammotheidae	Mobile deflexed	In juvenile rudimentary in adult	4–9	10 or less ♂, ♀	Condensed usually segmented	3,4	1,2,3,4
Tanystylidae		Achelate	4–6	10	4		
Pallenidae	Large fixed	Large scape 1—jointed	Absent or rudimentary	10 ♂, ♀	Well segmented	(1,2),3,4	1,2,3,4
Phoxichilidiidae	Large fixed	Large scape 1—jointed	0	5–6 ♂	Well segmented	1,2,3,4	1,2,3,4
Eudeidae	Large fixed	0	0	Simple 7 ♂	Well segmented	2,3,4	1,2,3,4
Pycnogonidae	Large fixed	0	0	Small 9 ♂	Segmented condensed	4	4

larger, and the femoro-tibial joint permits flexion in a ventral arc only. Flexion of tibia 2 is similarly confined to a ventral arc but the muscles responsible which are located in tibia 1 are approximately equal in size, and are rather longer and thinner than those of the more proximal leg segments.

The tarsus also flexes in a ventral arc and the flexor muscle is clearly evident, but this segment has no extensor muscle. The propodium articulates on the tarsus primarily in a dorso-ventral plane, and although containing the claw muscles it has no evident operative musculature of its own. Its movements during swimming are thus presumably entirely passive (Morgan, 1971). Each leg beats identically in a dorso-ventral plane, being extended during the downward power stroke and flexed during the mainly upward recovery stroke (Fig. 9d). Articulation of the leg joints occurs in a fixed sequence distally. At the start of the power stroke, as indicated by the start of the downward movement of the femur, tibia 1 is fully extended while the basal joint of tibia 2 is still opening out. Tibia 1 remains extended during most of the downstroke but starts to bend near the end of the stroke.

Assuming a fixed limb profile at constant angular velocity, maximum lift was calculated by Morgan (1971) to occur with the femur inclined to the dorso-ventral body axis at an angle of about 50°. At this position the outward component of the lateral thrust decreased to zero and when further declination of the femur occurs the lateral forces become inwardly directed. Of the different segments of the leg, tibia 2, the tarsus and propodium contribute most of the hydrodynamic force (Morgan, 1971). Thus when the limb is moving up and down in a high elevation beat in relation to the body the total upwards-directed lift is 45 per cent greater than when the limb moves through a low elevation beat. In the latter movement a considerable part of the final stages of the downward movement are directed laterally inwards and thus lose their effect as a lifting agent.

The degree of flexion which reaches its maximum during the recovery stroke and is greater when the leg beats through an arc of high elevation, so the leg radius of the legs is still considerably lower during the recovery than during the power stroke. This reduces drag on the legs during the recovery stroke (Fig. 9d).

Horizontal movement of the leg normally occurs at the start of the power stroke, with the femur in the elevated position (Fig. 9d). In animals crawling on the sea-bed the legs flex in a similar sequence to that used in swimming, and if the claw makes contact with a suitable substrate this flexion results in the animal being drawn towards the point of attachment and thus move forward. In many species, e.g. *Achelia echinata,* the propodus and terminal claw respond to tactile stimuli by apposing one another and so gripping the substrate. Without available anchorage for the claws, however, the leg-beat frequency varies considerably, and when striding slowly the movement of one or more of the legs is frequently arrested, apparently at random during the leg beat cycle, so that an interval of varying duration is evident between successive beats. Prior to the onset of swimming, however, the legs beat more consistently and the relationship between the frequency of the leg beat and the vertical swimming speed for non-reproductive *Nymphon* is shown in Fig. 9f. The lowest frequency recorded for a swimming animal was 0·7 beats/sec and at lower frequencies the animals started to sink slowly through the water (Morgan, 1971).

When the animal is sinking through the water the legs are extended horizontally and beat only occasionally, presumably to control descent, but in addition to this attitude *Nymphon* frequently adopts a 'plummeting' posture (Prell, 1910; Morgan, Nelson-Smith and Knight-Jones, 1964) with the legs folded dorsally (Fig. 9e). In this posture the rate of descent is more than doubled (Fig. 9f) (Morgan, 1971). On arrival at the substrate it often falls onto its side but when the legs are again spread it becomes correctly orientated with its dorsal side facing upwards.

When swimming, *Nymphon gracile* can move forwards or backwards. Presumably by changing the efficiency of the beat of some of the legs the body may either be angled at 45° to the horizontal, with the anterior end higher than the posterior which results in movement being backwards, or the body may be angled with the anterior end lower than the posterior with the resultant movement being forwards (Prell, 1910).

Anoplodactylus lentus which is considered to be a moderate swimmer can cover 12–15 centimetres in 30–40 seconds. *Callipallene brevirostris* moves more rapidly than this and *Nymphon mixtum* assisted perhaps by the long bristles on its tibia can attain 1–1·5 cm/sec (Prell, 1910). Prell related the swimming abilities of four sublittoral species of *Nymphon* to the lengths of their tibial bristles and described how in *N. mixtum* these bristles are outspread during the power stroke but pressed close to the leg during the recovery stroke. The swimming speeds of *N. gracile* (Fig. 9f), however, compare favourably with those reported for *N. mixtum* though in the former tibial spines are few and very short and Morgan (1971) was unable to determine whether they articulated during the leg beat as described in *N. mixtum*. Morgan (1971) stated that an increase in the rate of vertical ascent was generally achieved by increasing the leg beat frequency but occasionally unusually rapid upward excursions occur (Fig. 9f) suggesting that swimming animals have considerable lift force in reserve. This was not investigated further but would be especially significant for males carrying eggs on their ovigers and might well be achieved by increasing the elevation of the arc of movement of each leg during the swimming cycle (Morgan, 1971).

When it is free swimming, *N. gracile* is attracted to light over a wide spectrum but its response is more marked near the red part of the spectrum. This response possibly varies with age, sex or the time of year. If the animal is on a rock it does not respond to the light stimulus so readily. A few species such as *Pycnogonum littorale* move away from the light.

N. gracile has an endogenous tidal rhythm of swimming activity in the British Isles. In the winter and autumn months, the majority are active on the ebb (Isaac and Jarvis, in press). In this species there is a swimming response to pressure changes. A sudden increase in pressure tends to inhibit swimming, while a reduction promotes it. In the laboratory it showed no responses to cyclical pressure changes similar to those produced by onshore waves. Cyclical pressure changes approximately equal to tidal range and frequency produced activity equivalent to late ebb or low water. At some pressure frequencies a few individuals exhibited the opposite response which may be depth-regulatory (Morgan, Nelson-Smith and Knight-Jones, 1964). Little is known concerning the mechanisms of these responses and their uses to the animal can only be guessed at.

The pressure cycles which are most effective in evoking a depth-regulatory behaviour are those in which the rates of change ranged smoothly, every few minutes, from 0 to approximately 1 or 2 cm/sec (Fig. 10c). The rates of change most likely to be encountered during swimming are of this order. It has been suggested that this depth-regulatory response has a survival value. Probably it prevents individuals from swimming too far from their habitat, and would also allow economy in the use of energy, which may be critical when traversing wide areas of sand in search of small patches of suitable rocks (Morgan, Nelson-Smith and Knight-Jones, 1964).

N. gracile was attracted, in particularly large numbers during the winter months, to a light hung over the side of a boat at Concarneau (Fage, 1954) (Fig. 10b). A shore survey in the Bristol Channel showed that there were very few individuals present on the shore at this time although they occurred in reasonable numbers at other times of the year when few were collected from the plankton (Fig. 10a). This suggests that an on- and off-shore migration occurs. A variation in time of swimming in relation to the tidal cycle would effect this with minimum energy expenditure. There is little information concerning

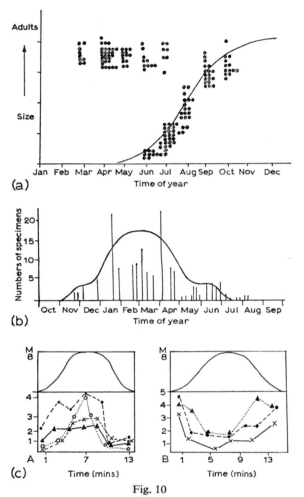

Fig. 10

(a) Graph showing the occurrence of individuals of all sizes of *Nymphon gracile* on the shore in the Bristol Channel.

(b) Graph showing the frequency of *N. gracile* in night hauls of plankton at Concarneau.

(c) (A) Depth-regulatory response shown by four individuals out of a group of ten *Nymphon* subjected to pressure cycles of 8m range and 14-minute period. Each line represents the mean activity during nine cycles of one individual recorded every two minutes. The pressure cycle above represents the nine cycles, superimposed on one another. (Redrawn from Morgan *et al.*, 1964)

(B) Response opposite in sense shown by three different individuals from the same group of ten.

movements of this type but it is probable that many species which are capable of swimming have this ability. The species which are either inefficient swimmers or cannot swim at all show indication of a similar phenomenon in that their numbers fall markedly during the winter months. Whether this is a result of movement off-shore or a high mortality which is made good in the spring by specimens washed up the shore has not been ascertained.

In *N. gracile,* it appears that only a minute proportion of the population swim at any one time. On a rocky bottom very brief and rare swimming periods would suffice to find new feeding grounds and to disperse the embryos. The capacity of some individuals to swim for hours which has been observed in the laboratory, with only occasional visits to the bottom to rest, would allow them to be carried for miles by tidal currents over sandy or otherwise unsuitable substrates. It seems likely that *Nymphon* can discriminate between tidal pressure cycles and more rapid changes, such as they would encounter during vertical swimming. A considerable amount of work is needed, however, before a fuller explanation of this species and others can be given.

Many authors have recorded encrusting organisms on pycnogonids. These epizooites include polyzoans, foraminiferans, hydroids, brachiopods, sponges, tunicates, serpulids, molluscs, stalked cirripedes and small anemones. It is presumably inevitable that slow-moving animals, with a firm exoskeleton, should serve as sites for such animals. Males of *N. gracile* become heavily encrusted during the period when they are carrying eggs on their ovigerous legs. This is presumably correlated with the fact that those appendages can no longer fulfil their preening and cleaning role when they have balls of eggs on them so that the pycnogonid has no way of cleaning its cuticle. It is not known whether the males die after the larvae are released, clean themselves or undergo an unusual moult. The second possibility seems unlikely since the ovigers would be unable to remove encrusting polyzoans (Gordon, 1932). The significance, from a distribution aspect, to the encrusting animals has not been determined though it has been suggested that pycnogonids aid the distribution of some species of molluscs in this way (Hedgpeth, 1964).

A number of pycnogonids have been observed swimming under water or in the plankton. Specimens of *Endeis spinosus, Pallene* and *Nymphon* have been collected at the surface or with a midwater net at the Tortugas laboratory, Florida. *Endeis mollis* Carpenter (Calman, 1923) has been taken swimming at the surface in the port of Nancowry, in the Nicobar Islands. The most numerous records of pycnogonids in the plankton have, however, been made in South Africa, mainly from Port Natal. The species recorded were *Nymphon bipunctatum, N. natalense, Pseudopallene, Gilchristi* and *Anoplodactylus pelagicus* (Flym, 1928). It is possible that there is normally no pelagic phase amongst pycnogonids and the records, which are most frequent in the summer months, result from hydroids or weed, to which the pycnogonids were attached, being torn from the bottom and bringing these animals up with them. Carpenter (1904) suggested that benthic forms, such as *Nymphon leptocheles* and *Anoplodactylus typhlops,* are sometimes captured in midwater trawls because some of the gear associated with the trawl may come into contact with and stir up the bottom.

Some species of *Colossendeis* and *Endeis* possibly speed their movements over the bottom by kicking themselves off the substratum though it is doubtful whether this mode of locomotion can be classified as swimming. Two species, *Nymphonella tapetis* and *Rhynchothorax philopsammum,* burrow in sand (Arita, 1937; Hedgpeth, 1951).

An important factor in the distribution of some small species in the Atlantic appears to be Sargassum weed. *Anoplodactylus petiolatus, Endeis spinosa* and *Tanystylum orbiculare* are species which have been most frequently found on weed in various localities, including some stations in mid-Atlantic. Most specimens are obtained from hydroids such as *Obelia dichotoma* which grows on the weed. Evidence supporting the importance of Sargassum weed in the distribution of pycnogonids is that at least nine species, all of small size and thus likely to be carried on the weed, occur on both sides of the Atlantic. Six have been recorded at Tortugas in the Florida Keys whilst on the eastern side of the Atlantic these species are scattered from Norway to Cape Verde. This pattern suggests a dispersal from the West Indies by the Gulf Stream (Fig. 35). Immature individuals of *Anoplodactylus petrolatus* have frequently been collected attached to medusae of *Obelia* sp. and of *Phialidium hemisphericum* in plankton nets off the south-west coast of Britain but the importance of this to the distribution of this species is not known.

The development of large numbers of very closely related species, so close in fact that there is often considerable dispute concerning their separation, has undoubtedly arisen, in part at least, because the pycnogonids, both as larvae and adults, frequently have very limited means of distribution.

If ambulatory limbs are broken off, the rupture usually occurs between the second and third segments of the limb. Pycnogonids have considerable powers of regeneration in the juvenile stages and when a break occurs, the regenerative process is very rapid. After the stub of the ruptured limb has sealed off, the new rudimentary limb is formed within the cuticle of the remaining part of the limb

External morphology   39

and is completed at a moult (Fig. 11d). This ability to regenerate is possessed by the immature stages and there is no evidence that adults can replace lost limbs. Regeneration of palps and ovigers occurs less frequently than of the ambulatory appendages. Regeneration involves not only the cuticle but also the digestive caeca and gonads, though many specimens only regenerate the digestive caeca and not the gonads. It is possible that limbs lost in a late larval stage regenerate in this fashion (Fage, 1949; Helfer and Schlottke, 1935). Regenerated limbs can usually be detected in the adult by their smaller size. Whether they eventually attain the full size has not been ascertained.

COLORATION

Pycnogonid species are usually white or have a characteristic colour but variations in coloration within a species are known to occur (Bouvier, 1923; Barnard, 1954; Lebour, 1947). Some species owe their colour to pigment in the body wall, but many, being transparent, show the colour of the epithelium or contents of the gut through the cuticle. *Phoxichilidium* is given its colour by pockets of the gut which contain unidentified purple symbionts (Helfer and Schlottke, 1935). Deep-sea pycnogonids usually have an orange–red coloration, and several examples of protective coloration are cited in the literature: *Lecythorhynchus marginatus, Tanystylum californicum* and *T. intermedium* all resemble *Aglaophenia* on which they live (Hedgpeth, 1951; Ricketts and Calvin, 1960; Stock, 1959). *Endeis spinosa* and *Anoplodactylus insignis* resemble *Obelia dichotoma* and *O. marginata* (Cole, 1904, 1910) respectively, and *Phoxichilidium virescens* and *Anoplodactylus angulatus* resemble *Ascophyllum* (Lebour, 1947) and *Corallinae*. These species are sometimes found on other organisms, where they become much more conspicuous. There is no evidence to suggest that pycnogonids can change colour readily and therefore the constancy of the coloration indicates a degree of long-term adaptation to a particular habitat to confer some protective coloration on pycnogonids. The pycnogonid *Anoplodactylus lentus* occurs on *Eudendrium* colonies which have a pale cream colour, and another pycnogonid, *Tanystylum orbiculare*, on a yellowish hydroid. Each of the pycnogonids is coloured so that it resembles the hydroid on which it normally lives. These species of pycnogonids are hardly ever found on the opposite hydroid but *Pallene brevirostris,* a whitish, almost transparent species, occurs on both hydroids and is thus presumably not adaptively coloured for either, unless perhaps the degree of transparency and its small size confer some protection.

Colossendeis colossea is claimed to be phosphorescent (Calman,

1923), a phenomenon which has also been observed once in *Nymphon gracile* (Jarvis and Isaac, 1972, Personal communication), during laboratory experiments. It is believed that the source of the emission in this species was the terminal segments of the first pair of legs, though as yet this has not been confirmed. The purpose of light emission in the pycnogonids is not known but, as in other animals living in a similar habitat, it may help to attract a mate, ward off a predator or transmit other messages to individuals of its own or other species. The group of cells or organ responsible for the emission has not yet been localised.

3

INTERNAL ANATOMY

INTEGUMENT

In common with other arthropods, the pycnogonids have an external cuticle which is hardened and fairly rigid over most of the body, but flexible at the joints to permit movement. With the use of various stains, both histological and histochemical, it can be demonstrated that this cuticle consists of a thin epicuticle, a hardened ectocuticle and a soft flexible endocuticle (Table 2). At the joints and other flexible regions the ectocuticle is absent (Fig. 11a).

There are a number of different processes by which the integument of arthropods becomes hardened. In insects, the hardening of the cuticle results from phenolic tanning and in crustaceans, although the cuticle is initially partially hardened by this process, the prime agent is calcification. In arachnids such as *Limulus* and *Palamneus* there is evidence that hardening of the integument results from the formation of sulphur linkages. Although the complete process is not yet understood, disulphide linkages are involved in the hardening of the cuticle in the Indian pycnogonid *Propallene kempi,* the only species so far examined. The cuticle of this species consists of an inner layer, characterised by horizontal lamellae and which is presumably equivalent to the endocuticle of the insects. Outside this there is a thinner apparently homogenous layer, presumably equivalent to the ectocuticle of insects which is bounded by a very thin layer which, in turn, is presumably equivalent to the epicuticle of insects (Krishnan, 1955).

The disulphide bonds have been demonstrated in the epicuticle by the fact that the epicuticle will stain with Mallory's triple stain after it has been treated with alkaline sodium sulphide, which breaks up the sulphide bonds and allows staining to occur (Wyer, 1972).

The cuticle is variously covered with pores, setae which may be

Table 2

Analysis of the composition of the cuticle of *Propallene kempi*
(Adapted from Krishnan, 1955)

Histological or histochemical stain used	Results Endocuticle	Epicuticle
Mallory's	blue	unstained
Heidenhain's	faint brown	unstained
Schultz	chitin present	chitin absent
Chitosan	—	—
Millon's	—	—
Xanthoproteic	—	—
Argentaffin	—	— (outer layer positive)
Sudan Black	—	—
Thioglycollate	—	+ } suggest epicuticle hardened by disulphide bonds
Alkaline lead acetate	—	+

+ = positive result
− = negative result

very long as in *Eurycide* or *Ascorhyncus*, tubercles and spines. In some species spines are larger in the males than in the females. This is particularly marked in species of the genus *Achelia* such as *A. echinata* and *A. longipes*. There is a slight tendency for species which

Fig. 11
(a) Section through the cuticle of *Nymphon gracile* drawn from an electron micrograph.
(b) The pieces into which the cuticle of *N. gracile* splits during moulting.
(c) The femur of a male *Endeis spinosus* showing the gland which is believed to be responsible for the cementing of the eggs to the oviger of the male and one of the ducts from this gland. The form of the duct is shown in *Ammothella appendiculata* where it opens at the end of a spine and in *Anoplodactylus typhlops* in which the spine is shorter.
(d) Stages in the regeneration of a leg of *N. gracile*.

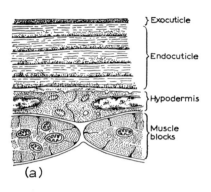

(a)

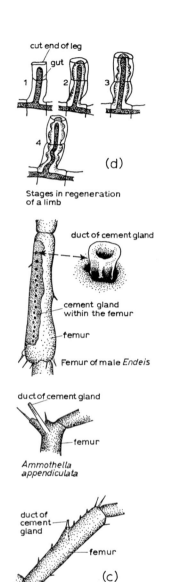

(d) Stages in regeneration of a limb

Femur of male *Endeis*

Ammothella appendiculata

Anoplodactylus typhlops

(c) Cement glands and ducts

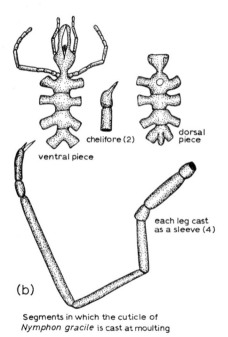

(b) Segments in which the cuticle of *Nymphon gracile* is cast at moulting

Fig. 11

live in deeper water to be less spiny than littoral relatives but there are many exceptions to this, e.g. *Nymphon hirtum*.

The cuticle is traversed by the ducts of numerous cutaneous glands which develop in the hypodermal cells over much of the body. Some of these contain grains of what is thought to be excretory material. Some of the ducts have a sphincter at the outer end which permits closure when necessary (Fig. 17e). The pores, which have a similar appearance in most species, are particularly abundant and are relatively larger in *Pycnogonum littorale*. As in other arthropods growth necessitates moulting but this process has not been investigated in detail. The cuticle is shed in a number of separate pieces (Fig. 11b). In this species they do not occur on the distal third of the proboscis, on the lens part of the eyes or the ventral surface of the trunk, but elsewhere they are abundant. Many of these ducts lead from cells or groups of cells which, when viewed under the electron microscope, have the appearance typical of mucus-secreting cells (Storch and Welsch, 1972) (Fig. 17e). The function of the secretion of these cells is unknown, they may help or hinder desiccation when littoral species are exposed by the falling tide. Under some conditions the secretion may help respiration to occur over the body surface or, since many of these animals feed on actinians or hydroids, the secretion may protect the soft regions from the nematocysts of their host. The cuticles of annelids and tardigrades have been shown to have an outer layer of mucus and this similarity with the pycnogonids, if not purely an adaptation to the environment, may lend tenuous evidence for a consideration of affinities (Baccetti and Rosati, 1971).

In *Pycnogonum littorale* the cuticle of each leg is shed as a sleeve and the cuticle of the body splits longitudinally on each side of the body, so that it is shed as a dorsal and ventral piece (Lotz, Von Guntram and Bückmann, 1968). In *Nymphon gracile* the process is similar but this species has chelifores which the previous species lacked. In *N. gracile* the cuticle of each leg is cast as a sleeve and the trunk cuticle with lateral breaking lines is shed as a dorsal and ventral unit (Fig. 11b). The dorsal piece includes the eye tubercle, the covering of the eyes and the whole abdomen surface together with the lining of the hindgut. The ventral section includes the whole covering of the proboscis and the lining of the foregut, including the mesh of spines forming the sieve. The whole covering of the palps and ovigerous legs remains in contact with the ventral section. In common with a few other arthropods such as some members of the Crustacea, *P. littorale* continues to moult after it has reached sexual maturity (Fig. 20a). It is likely that other species might also have this characteristic.

GLANDS

In all pycnogonids investigated, there are glands in some or all of the appendages with ducts through the cuticle to the outside. Within the larval chelifores but not in those of the adult there are glands consisting of a number of flask-shaped cells (Fig. 25a). These cells are situated in the basal segment of the appendage with a duct from them opening at the distal end of the chelifore. Sometimes the duct opens at the tip of a long, mobile spine, as in some species of the genera *Ammothea, Pallene, Tanystylum* and *Nymphon*. This gland secretes a sticky thread by which the larva attaches itself to other larvae, the ovigers of the male parent and eventually to the host. In *Nymphon hamatum*, several filaments are secreted by separate sacculi of the gland. In *Pycnogonum littorale* the spine, at the tip of which the duct opens, is a long, thin filament. There are glands present in the larval chelifores of members of the genus *Phoxichilidium*, but there is no long spine or tubercle and the secretion produced by the glands is emitted from many small orifices set along the opposing edges of the chelae. There is considerable variation in the degree of development of the spine even within closely related species. In the larvae of *Ascorhynchus arenicola* the spine is absent, but in a related species, *Ascorhynchus castelli*, it is extremely long— more than twice the length of the body. In some other species of the genus *Ascorhynchus* and in some members of the Nymphonidae, e.g. *Boreonymphon robustum*, the larvae have reached a more advanced stage of development at the time of hatching compared with those of other pycnogonids and there are no glands present in the chelifores.

Vesicular-shaped glands have been described in the fourth and fifth segments of the palps and the third and fourth segments of the ovigerous legs. They are present in the elongated third segment of the palps of *Tanystylum* and open by a sieve-plate at the end of the second joint. In *Ammothea* and *Ascorhynchus* these glands open through a small tubercle on the fifth joint of the palp and they are also present in larvae of the Ammotheidae even before these have attained their full complement of legs. No function has been suggested for these glands. The males of many species have another type of gland in their femora which are believed to act as cement glands. The secretion produced by these glands sticks first to the eggs and then to the ovigerous legs of the young larvae and in some examples completely coats the egg mass. In most species, the glands open by a single orifice which may be situated at the tip of a spine or by a few pores grouped closely together (Fig. 11c). In other species the openings are distributed over a wide area as in *Ascorhynchus arenicola* and *Endeis spinosa*. In the latter species a row of pores occur along the whole length of the male femur (Fig. 11c). In most species the ducts appear for the first time at the last moult before maturity is attained (Thompson, 1909).

Mucus glands have already been mentioned in the section dealing with the cuticle and there is an unconfirmed suggestion that some cuticular glands produce a defensive secretion.

ALIMENTARY CANAL

The alimentary canal of the adult pycnogonids consists of a number of distinct regions.

At the tip of the proboscis is situated the mouth, which leads into a pharynx passing down the centre of the proboscis and at the anterior end of the cephalic segment the pharynx leads into the oesophagus which in turn opens into the mid-intestine. The junction between the oesophagus and mid-intestine has a valve-like constriction. The mid-intestine extends throughout the whole body and sends blind diverticulae into the appendages. The mid-intestine eventually joins the hindgut which lies within the abdomen and opens to the outside by a terminal anus through a triangular valve (Helfer and Schlottke, 1935) (Fig. 4a). The pharnyx is the suction organ and masticatory apparatus lying within the proboscis (see page 11).

The oesophagus, which is situated in the first body segment, is lined with a thin chitinous intima but there are no associated spines or 'teeth'. Beneath the cuticle is a hypodermis, composed, near the anterior end, of a single layer of cells, each of which has a large nucleus. In contrast to the structure of the pharynx, there are no conspicuous muscle layers associated with the oesophagus. In the posterior part, the wall of the oesophagus becomes multilayered and the cuticle becomes progressively thinner until it disappears. In this region within the cytoplasm of the gut cells numerous proteinaceous globules are present and it has been suggested on histological evidence that there is a gland in this region which functions in a manner similar to the pancreas.

At its posterior end, the oesophagus projects slightly into the midgut and a valve is present at this point which prevents regurgitation of food pulp from the midgut. This is important since anterior to this point the injected food is moved backwards and forwards. The structure of this region resembles the 'valvula' in the midgut of many arthropods and the pressure of the food in the midgut forces the flaps of the 'valvula' together thus occluding the lumen of the oesophagus (Helfer and Schlottke, 1935; Fage, 1949) (Fig. 4a). The valvula is presumably opened when food in the midgut has either

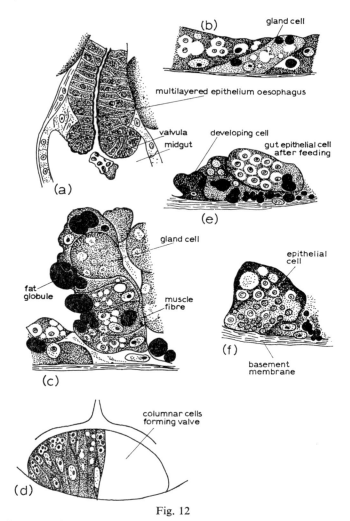

Fig. 12

Structure of the gut cells and changes in it during a feeding cycle.
(a) Structure of the epithelium at the junction of the oesophagus and midgut.
(b) Structure of normal gut epithelium.
(c) The formation of a villus.
(d) The structure of the valve between the mid- and hindgut.
(e) Appearance of a gut epithelial cell after feeding.
(f) Secondary gut epithelium.

been absorbed or voided whilst pressure builds up within the oesophagus during feeding and food can then pass into the midgut once again. The midgut, which is the site of the majority of digestion in the alimentary canal, extends throughout the rest of the trunk with the diverticulae passing from it. Sometimes there are additional caeca into the proboscis, one pair in the proboscis of *Nymphon gracile* and *Phoxichilidium femoratum* and in *Endeis spinosa* there are two pairs. The midgut epithelium consists of three types of cell: gland cells, digestive cells and developing cells (Fig. 12b,c). The last give rise to the other two types. The gut cells, at least in the early stages of development, have smooth muscles at the base, the fibres of which spread out and interweave. Maceration of the tissues of the gut shows isolated epithelial muscle cells with processes similar to those found in coelenterates such as *Hydra*. The developing cells of the midgut are initially almost spindle-shaped and approximate to the basal membrane but during their transformation to a gland or digestive cell (Fig. 12c), the cell body enlarges towards the lumen of the gut. This enlargement results from the production of secretory globules or absorption of food, depending upon which type of cell is involved.

Digestion is intracellular and, as food passes down the gut, the basal muscles contract and the free ends of the villus-like cells expand outwards. Some of these become detached from the rest during peristalsis, and can wander about in the gut lumen either singly or in groups. The connections between individual cells become dissolved but the cells can later re-attach to the cells of the gut epithelium. During their wandering in the gut lumen and later their secondary attachment to the gut epithelium, digestion occurs within them. The waste products of this process remain embedded in these cells. After the wandering phase and when digestion and secondary attachment have taken place, only globules of high refractive index and slightly stainable masses remain in these cells. These inclusions are presumably the residue of the protein digestion which has occured (Fig. 12e).

The re-attached cells, or secondary gut epithelium, differ from the cells of the original gut epithelium in that there is no basal membrane or associated muscle fibres (Fig. 12f).

Eventually the cells become free from the epithelium again and become involved in the peristaltic flow of the gut contents into the hindgut from which they are eventually expelled through the anus. During this phase they function as excretory cells, still containing the residues of digestion.

The number of gut caeca varies with the number of appendages. There is often a branch of the gut in the chelifores as in *Nymphon gracile* and *Phoxichilidium femoratum,* but in some species this branch is only poorly developed. In *Nymphon gracile* the caecum reaches almost to the claw of the chelifore.

In some species of pycnogonid the body is greatly shortened and the organs of the anterior segments are displaced. This shortening is very conspicuous in the gut of *Colossendeis*, in which the midgut is much shortened and all the caeca arise from the level of the first walking leg. They then pass posteriorly alongside the gut until they curve off into the legs. In *Phoxichilidium femoratum* and *Pycnogonum littorale* the gut caeca of the appendages reach as far as the sixth joint, which is the second tibia, while in *Nymphon gracile, Nymphon rubrum, Achelia echinata* and *Endeis spinosa* the caeca reach the propodus.

The lumen of the gut caecum narrows towards the distal end of the appendage until a solid cord of cells remains. This finally narrows to a single cell, which is no longer completely functional; the nucleus of this type of cell is usually pycnotic. The structure of the gut caeca is basically similar to that of the rest of the midgut. The ends of the caeca are attached to the body wall by files of connective tissue which are connected to flat cells in the distal termination of the gut epithelium. A sheet of connective tissue surrounds the gut and mesenteries of the body and appendages are attached to the epithelium of the gut and thus, by partially dividing up the haemocoele, they help to regulate the circulation of the blood (see page 50) (Helfer and Schlottke, 1935).

The midgut opens into the hindgut which is situated in the caudal or anal segment, often referred to as the abdomen. The passage from the midgut to the hindgut is through a valve, which has a similar structure to the one between the oesophagus and the midgut. The valve in cross-section consists of three units, a single dorsal one and one or two ventral ones. The position of the valve means that excretory cells within the hindgut can only be driven in one direction, towards the anus. The cells of the valve are columnar in shape and contain numerous protein globules at their distal ends but the function of these has not been determined. It is possible that some form of lubricant is produced to protect the cells of the valve. In the middle section of the hindgut the epithelial cells are less columnar in shape and consequently the lumen is larger (Fig. 12d).

There are cells in the anterior part of the hindgut which contain protein residue and possibly have an excretory function but they are absent from the more posterior region. The hindgut ends with a perpendicularly-orientated anal opening or slit. There are oblique muscles, which run forwards from the walls of the anal appendage, and attach to the posterior section of the hindgut. When these muscles contract, the anus opens. The stimulus for contraction of these muscles is probably the filling of the hindgut (Helfer and Schlottke, 1935).

CIRCULATORY SYSTEM

The circulatory system of the pycnogonids is of the open type. The blood, which in common with that of other arthropods is called haemolymph, circulates throughout the body (Fig. 13a). This movement of haemolymph is brought about by a median vessel in the dorsal part of the cephalothorax. In *Endeis* this vessel extends from a point below the eye tubercle to the base of the abdomen. In this genus and in *Nymphon rubrum* there are two pairs of lateral valvular openings or ostia, one opposite the second and the other opposite the third pair of ambulatory appendages (Fig. 13a). The walls of the vessel are muscular, but these muscular walls do not extend around the vessel dorsally. In this region, its lumen is only covered by the hypodermis and cuticle of the back (Fig. 13b). The cephalothorax and appendages have a number of septa, which divide them into compartments which aid the circulation of the haemolymph. The most important of these septa is the horizontal one which lies between the dorsal vessel and the gut and above the gut in the appendages. This septum is perforated, in the region of the lateral processes, by slits, which place the dorsal and ventral cavities of the haemocoele in communication. In the ambulatory appendages, the septum does not reach to the end of the limb and it is claimed to be capable of movement possibly to aid the circulation of the haemolymph.

The course of the circulation is generally outwards in the inferior or ventral sinus and inwards towards the heart in the superior or dorsal sinus. This direction of circulation also occurs in the appendages but in the proboscis contraction of the dorsal vessel drives the haemolymph forwards, away from the anterior end of the dorsal vessel, in the dorsal channel. Contraction of the dorsal vessel occurs two or three times each second in *Phoxichilidium* species, but in species with small bodies and long legs, the circulation of the haemolymph is actuated more by movements of the limbs and contractions of the intestinal caeca rather than by the direct effect of contractions of the dorsal vessel (Helfer and Schlottke, 1935).

Specimens of *N. gracile* have been observed moving the body up and down in the water by flexing the legs. This movement probably aids circulation of the haemolymph and in addition may aid respiration though whether this movement occurs only at times of respiratory stress has not been determined.

The haemolymph has several functions in the pycnogonids. First

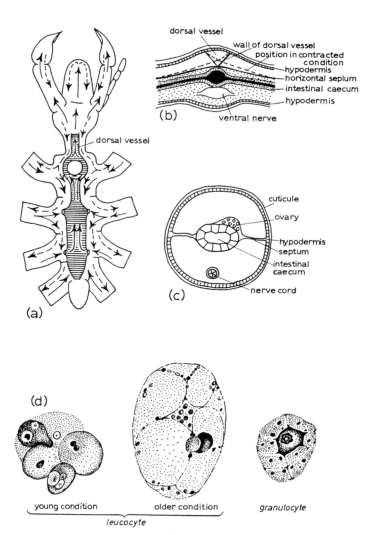

Fig. 13
(a) The position of the dorsal vessel in the body, the positions of the ostia and the direction of flow of the haemolymph within the dorsal vessel and the haemocoele.
(b) Transverse section through the body. (After Cole, 1901)
(c) Transverse section through the basal joint of the third leg in *Phoxichilus charybdaeus*. (After Dohrn, 1881)
(d) Types of cells within the haemolymph.

it bathes the organs of the body and probably transports nutriment and hormones around the body dissolved in it. Second, within conical-shaped spaces in the cuticle, which are only separated from the outside by a thin membrane, blood cells have been observed and it is possible that respiration occurs by gaseous exchange across these thin membranes though conclusive evidence is lacking. If this does occur, however, the haemolymph presumably acts as a transport system for oxygen and carbon dioxide. The possible role of the haemolymph in the transfer of pressure from one part of the body to another during ecdysis has not been investigated, but it is likely that this transfer occurs in order to split the old cuticle.

The haemolymph is a colourless plasma which contains several types of corpuscles (Fig. 13d). In the early stage of development of the pycnogonid larva there are hyaline leucocytes and later granulocytes in the haemolymph. It is possible that the latter type give rise in the adult to amoeboid cells which are present in the haemolymph at that stage and which are highly active, frequently coalescing to form plasmodia. These are one of several cell types present in the haemolymph of the adult though little is known beyond a very brief histological description and no work has been done to date on their respective functions (Sanchez, 1959).

The dorsal vessel appears relatively late in embryological development when the larva has three pairs of legs (Dawydoff, 1928).

EXCRETION

The absence of a coelom, which most workers believe has completely regressed, is accompanied by an absence of discrete excretory organs. In other animals these are usually formed from expansions of the coelomic cavity.

In the larvae, the elimination of waste products by the proctodeum is considered unlikely because it does not join the rest of the gut until the end of larval development. A relatively large cell has been described (Thompson, 1909) at the base of each chelifore in the protonymphon larva which resembles a nephrocyte with a typical inclusion (Fig. 25). It is possible that the material which is deposited in this cell represents the waste products of metabolism. It is more probable, however, that during larval life, excretion occurs through the body surface mainly at the time moulting occurs and that the waste products awaiting release at ecdysis are deposited in vacuoles in the hypodermis. Immediately before moulting, the hypodermis of a larval *Achelia* is literally packed with these products, but, after the moult, the hypodermis is devoid of these substances. In the adult, the absence of nephridia and Malpighian tubules suggests that waste

Internal anatomy

products are either eliminated by the gut when cells detach from the epithelium or are passed out from the epidermis and it is possible that both these sites contribute to the process (Sanchez, 1959), in forms which continue to moult in the adult stage.

The results from the use of dyes in a number of species suggest that areas of the body surface which are involved in excretion lie on the lateral projections at the bases of the legs (Fig. 17c).

NERVOUS SYSTEM (FIG. 14)

The nervous system in pycnogonids consists of a supra-oesophageal ganglionic mass or brain which is connected by circumoesophageal commissures, to a chain of paired ventral ganglia. The brain is believed to consist of a protocerebrum and tritocerebrum only (Fig. 14b,c). In the trunk of *Nymphon pixellae* at the level of the first pair of walking legs there are situated dorsally four optic nerves (Fig. 14b,c) (Bullock and Horridge, 1965), cheliforal nerves arising from the front of the brain together with a median rostral nerve and paired stomodeal nerves which go to the proboscis. Within the proboscis, these nerves are joined to each other by ladder-like connections at intervals along the organ. In addition to these nerves, a paired nerve which arises from the suboesophageal ganglion runs to the dorsal and another to the ventral muscles of the proboscis. Some workers have described a ganglion near the tip of the proboscis but this is possibly only a group of sensory cells (Fry, 1965).

The ventral ganglia correspond to the appendages in number and position. Inter-ganglionic connections are long in elongated species such as *Nymphon*, but in *Tanystylum* and *Ammothea*, which are short-bodied, the ganglia of adjoining segments touch each other and the connectives are insignificant. In each segment there arise two nerves to each leg and a paired nerve to the ventral muscles of the segment (Fig. 14a). One leg nerve serves the dorsal part, above the horizontal septum, and the other the ventral part. The terminal trunk ganglion sends three paired nerves to the abdomen.

Little is known regarding the structure of the brain. The central body of the brain is situated ventrally in relation to the optic tracts. Adjacent to the optic nerve on each side there is a frontal organ which is connected by a nerve to the optic tract. In addition to these there are small frontal organs, perhaps sensory in function, which have a nerve supply to the same root. There are no corpora pedunculata—nuclei associated with the brain of annelids and in the protocerebrum of mandibulate arthropods. The position of these nuclei in the hind part of the annelid brain, which may have evolved into the deutocerebrum of the arthropods, may account for their absence in a group which lacks a deutocerebrum. In other offshoots from the annelid stock the corpora pedunculata may have been displaced into the protocerebrum prior to the loss of the deutocerebrum. In crustaceans and insects the deutocerebrum contains nuclei associated with the first antennae of the former and the antennae of the latter and gives off nerves to these but in the chelicerates and pycnogonids there are no appendages corresponding to the first antennae of crustaceans and the antennae of insects, and the deutocerebrum has disappeared. The protonymphon larva has one pair of eyes which later produce another pair and this accounts for their being only two optic nerves (Bullock and Horridge, 1965) (Fig. 20).

ENDOCRINE SYSTEM

Most of the nerve cells, especially those in the brain and ventral ganglia, are medium-sized. In certain zones, however, large cells occur which contain secretory material usually in one or perhaps more deeply staining spherical droplets (Fig. 14d,e,f). In some cells there is a large secretory inclusion, a large nucleus and a thin peripheral band of cytoplasm. A comparison of the large cells in the nervous system of *Phoxichilidium femoratum* and *Endeis spinosa* with neurosecretory cells in a number of vertebrates and invertebrates suggests that this may be their function in pycnogonids (Hanstrom, 1965).

In some pycnogonids an organ has been described situated beneath the eyes laterally to the optic lobes of the brain. It is surrounded by a thin layer of connective tissue and consists of a large number of cells which resemble unipolar nerve cells. Distally, each of these cells has a dumb-bell-shaped body which stains intensely with a number of stains and is proximally attenuated into fibres. This organ has been called Sokolow's organ (Hanstrom, 1965) (Fig. 17a). It has been suggested that it represents specialised nerve cells which are the

---

Fig. 14

Diagrams of the nervous system in *Nymphon pixellae*. The position of nerves entering the brain, the structure of typical neurosecretory cells and the ventral organ are shown.
(a) Dorsal aspect of *Nymphon pixellae*.
(b) Lateral aspect of the cephalic nervous system of *N. pixellae*.
(c) Ventral aspect of the cephalic nervous system of *N. pixellae*.
(d) Ventral organs in the young larva of *Chaetonymphon*, ganglia still connected to the organ.
(e) Older larva than depicted in (d) with ganglia separated from the ventral organs.
(f) Neurosecretory cells amongst the nerve cells in the first ventral ganglion.

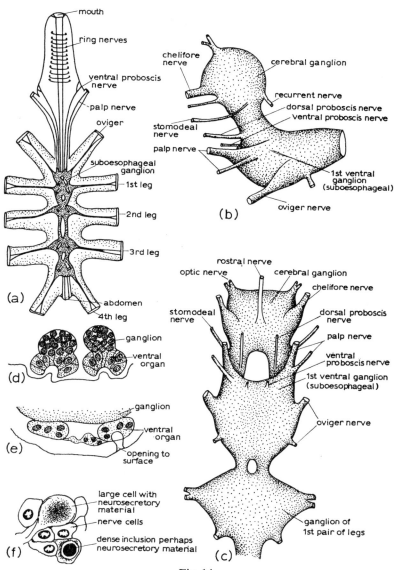

Fig. 14

pycnogonid counterpart of the corpora cardiaca or corpora allata of insects or the X-organ and sinus gland of crustaceans. Its function has yet to be elucidated, however, together with the rest of the endocrine system of the pycnogonids. Considering the importance of this system in other arthropods, it is hardly likely that the pycnogonids lack one.

In some pycnogonids, the differentiation of the nerve ganglia during embryological development is accompanied by the formation of a series of paired, segmentally arranged structures, usually referred to as ventral organs (Fig. 14d,e). Each pair of ganglia, including those which form the protocerebrum, corresponds to a pair of these organs. They are produced by invaginations of the ectoderm which extend inwards as far as the ganglia. The ventral organs sometimes have a cavity but usually they are present as a compact mass of large cells. They usually disappear during the larval period. It is thought that they are homologous to similar structures in *Peripatus* and *Scolopendra*. A number of functions have been suggested for them. It has been suggested that they represent the beginnings of neuroblasts destined for the nerve ganglia, or neurosecretory structures, though when there is a lumen it opens to the outside, making this latter hard to accept (Dawydoff, 1928; Hanstrom, 1965; Sanchez, 1959).

SENSE ORGANS

The only sense organs in pycnogonids to which a function can be attributed with certainty are the eyes. These are on the lateral surfaces of the ocular tubercle, which is situated on the cephalon or first segment of the body (Fig. 1). The shape and position of the cephalon varies between species (Fig. 15a–i). In species such as *Ammothella appendiculata* the tubercle is relatively tall and in others, e.g. *Pycnogonum planum,* very short. In some, e.g. *Parapallene capillata* or *Ascorhynchus anchenicum,* it is situated well back on the cephalon whilst in others, e.g. *Pycnogonum planum,* it is near the front and may actually overhang the cephalon anteriorly as in *Austrodecus simulans, A. frigorifugum* and *A. breviceps.* This condition is more marked in *A. simulans* than in the other two examples (Fig. 15a). The size of the tubercle may vary between sexes. That of the male is taller and with a more pronounced knob on the top than that of the female in the New Zealand pycnogonid *Achelia variabilis* (Fig. 15i). The juvenile eye tubercle has a sharply pointed knob on top in *Tanystylum excuratum* which becomes lower and bluntly rounded on top in the adult (Fig. 15h).

There are usually four eyes and either all are of equal size or the

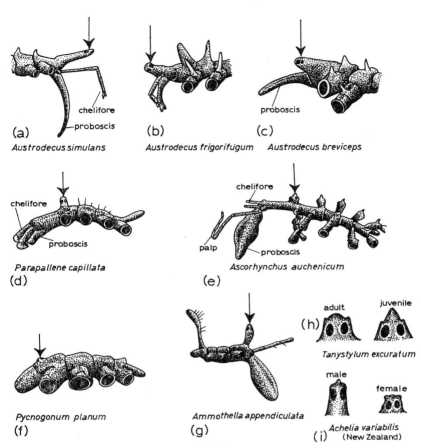

Fig. 15

(a–g) Diagrams of examples of pycnogonid species having eye tubercles of different shape and position.
(h) Difference between the shapes of the eye tubercles in adults and juveniles of *Tanystylum excuratum*.
(i) Difference in shape between the eye tubercles of male and female *Achelia variabilis*.

anterior pair are larger than the posterior pair. The larvae have one pair of eyes but during development another pair appear using the same nerve supply but their exact origin has not been determined (Fig. 16b). This usually occurs at the time in development when the protonymphon larva becomes the four-legged larva. The second pair of eyes are visible initially as a small patch of pigment. Pycnogonids living in water deeper than four or five hundred fathoms frequently have imperfectly developed eyes which may either lack a lens, masking pigment or, in one or two exceptional species, the complete eye (Hoek, 1881). This statement is a generalisation which is perhaps over-simplified since all species at these depths do not have reduced eyes and, at a given depth, some species have reduced eyes whilst others have normal ones. All that can be said safely is that there is a trend towards a reduction of the eyes in some way at these depths. Even within a single species such as *Nymphon stylops* there is a variation in the degree of development of the eyes at some depths (Hoek, 1881). *Ascorhynchus minutus* differs from the usual pattern in having reduced eyes although it is a shallow water species (Hoek, 1881).

The eyes of arthropods are either (a) compound eyes, composed of numbers of neighbouring units or ommatidia, each of which consists of 6–8 receptor cells, characteristic of insects and Crustacea, or (b) simple eyes which consist of a cup-shaped retina, a single lens and a number of photo-receptive cells between. This second type of eye is typical of the chelicerates, the ocelli of insects and pycnogonids. *Limulus,* the king crab, is unusual amongst chelicerates in having both types of eye but the compound eyes in this species differ in structure from those in insects or crustaceans and have possibly been evolved separately. The pycnogonid eye resembles that of arachnids (Hoek, 1881; Morgan, 1891; Sokolow, 1911; Wiren, 1918; Helfer and Schlottke, 1935; and Jarvis and King, 1972) and is situated on the lateral surfaces of the tubercle. Although the lenses of the four eyes are not joined, they cover most of the lateral surface of the ocular tubercle, thus giving good multi-directional light sensitivity except from below the animal (Fig. 16a).

The eye consists of an oval semitranslucent biconvex lens which is formed from modified cuticle with, on its inner surface, a vertical

---

Fig. 16
(a) Diagram showing the eye tubercle with the position of the eyes, optic nerves and brain.
(b) Vertical section through eye.
(c) Horizontal section through eye.
(b) and (c) have inset diagrams showing the ultrastructure of the different types of cell in the eye.

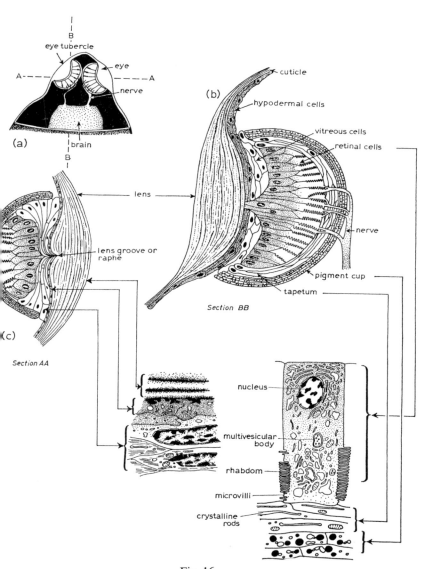

Fig. 16

groove (Fage, 1949) but without setae or pores on its outer surface (Fig. 16b,c). In *Nymphon gracile* the cuticle covering the ocular tubercle is 7·5–10μ thick but the lenses are expanded to 25–35μ. The cuticle of the ocular tubercle in section has 15–20 layers each with four lamellae and is secreted by the layer of hypodermal cells or cuticular epithelium which lie beneath it. The cuticular epithelium extends under the peripheral parts of the lens where it functions as the lens-secreting tissue (Fig. 16b,c). The cells secreting the lens are morphologically similar to those secreting the cuticle except that they have greater secretory activity. The secretion which forms the cuticle is contained in vesicles within the cells before release. The vesicles in the cuticle-secreting cells are 3–4μ in diameter whilst those in the lens-secreting cells are 12μ. In both types of cell the secretory droplets concentrate directly beneath the cuticle or lens and later fuse with it. The greater thickness of the lens can be partly attributed to extra layers of enticular material in between the other layers of the lens (Fig. 16b,c). The lens is not perfectly biconvex since most of its increased thickness is due to hypersecretion on the inside which causes the inner surface of the lens to bulge into the eye with a more marked curvature than is evident on the outer surface (Fig. 16b,c).

The hypodermal cells are flat with relatively large nuclei, electron-dense cytoplasm resulting from the numerous free ribosomes present in it, numerous vesicles and a few mitochondria which are relatively large with numerous cristae. Electron-dense granular material is abundant in the nucleus aggregated at the periphery of the nucleus and there is some golgi activity (Fig. 16b,c). All these characters support the evidence that these are actively secreting cells stated earlier (Jarvis and King, 1972b). Directly beneath the lens-secreting cells there are 'glassy' or vitreous cells which constitute the vitreous body (Fig. 16b,c). The vitreous body was first described in pycnogonids by Grenacher (1879) and Graber (1880). The vitreous cells are probably formed from modified hypodermal cells (Grenacher, 1879) but although they resemble these cells in size and shape their ultrastructure is quite different (Jarvis and King, 1972q) .Their cytoplasm contains few ribosomes, and those present are clumped together, and few vesicles and mitochondria, which, when present, are small and spherical with only one or two cristae, all of which suggests that these cells have only modest energy requirements and probably do not synthesise any substance in large quantities. The constitution of the cytoplasm and its organelles means that the vitreous cells are fairly translucent and thus provide less of a barrier to light reaching the retinal cells than would those of other hypodermal cells (Wiren, 1918). The vitreous body has been likened to a lens with the cuticular thickening acting as a cornea (Sokolow, 1911).

Internal anatomy

In the peripheral region of the eye cup the vitreous cells are adjacent to cells which have reflecting properties and collectively form the tapetum. The structure of these cells is similar to that of the vitreous cells and they have been regarded as the same tissue (Wiren, 1918) originating from hypodermal cells. The tapetum consists of a series of layers of cells each of which contains rod-like structures $1 \cdot 5$–$3 \cdot 0 \mu$ long and $0 \cdot 08 \mu$ wide which are thought to be crystalline structures (Jarvis and King, 1972b). These are responsible for the light-reflecting properties of these cells. Any light entering the eye and not immediately impingeing on a photosensitive area is reflected and may then do so. This is especially useful under conditions of low light intensity.

The retinal cells form a single layer on the inner side of the tapetum (Fig. 16b,c). Hoek (1881) considered that the retina of pycnogonids consists of rod-forming elements and ganglion cells. Studies on the ultrastructure of these cells in the eyes of *Nymphon gracile* have revealed that at the proximal end of the cell the membrane is elaborated into many interwoven microvilli which together with those of adjoining cells constitute a rhabdom. This structure was originally described as a cuticular secretion and later as a concentration of neurofibrils. It is now believed from evidence of the retinal cells of other animals that they are the actual photoreceptor structure and their large surface area and arrangement ensures an ordered plane arrangement of the visual pigment molecules for the most effective absorption of radiant energy (Jarvis and King, 1972b). The retinal cells are columnar in shape and around the periphery of the eye they are separated from the lens by lens-secreting cells and vitreous cells (Fig. 16b,c) but in the centre their distal ends, where the nuclei are situated, reach almost to the lens (Fig. 16b,c). The eye of *N. gracile* is of the non-inverted, prebascillar type in which the nuclear parts of the receptor cells are nearer to the lens than the rhabdoms (Fig. 16b,c). Thus light entering the eye passes through the nuclear part of the receptor cell before impinging on the rhabdoms. These nuclear areas of the retinal cells are synonymous with ganglion cells described by Hoek (1881). There are few mitochondria in the retinal cells with few cristae but there are numbers of vesicles, near to the rhabdoms and multivesicular bodies, which are $1 \cdot 5 \mu$ in diameter and contain numerous small electron-dense vesicles. These have been recorded in the retinal cells of other animals but their significance is not known (White, 1967). The cytoplasm contains a closely packed membrane system consisting of flattened cisternae (Fig. 16b,c). The pigment layer consists of 3–4 layers of cells each about 3μ long and 1μ deep when viewed in section (Fig. 16b,c). No nucleus has been observed in any of them and the cytoplasm appears empty except for a number of vesicles, some of which contain the screening pigment. It is possible that some movement of pigment occurs within the system of vesicles (Jarvis and King, 1972b). In some species an extension of the cuticle surrounding the pigment cup has been described (Hoek, 1881) but this is not present around the eyes of *N. gracile* (Jarvis and King, 1972b).

Each of the retinal cells is a sensory neuron and extends proximally as a nerve axon. The axon of each receptor cell leaves the eye independently of the others by passing through the tapetal and pigment layers. These axons collect behind the eye to form the ocellar nerve (Fig. 16b,c) (Jarvis and King, 1972b).

When the pycnogonid moults, the lens is shed but no information is available concerning the functioning of the eye at this time.

Presumably the eyes of pycnogonids can only detect light intensity and cannot form an image. Wavelengths extending into the blue and red can be detected but the eyes of *N. gracile* are more sensitive at the red end of the spectrum. More detailed investigation of this is needed but a similar phenomenon has been observed in other arthropods (Jarvis, 1972).

The pycnogonids have a number of other structures on their cuticles which are presumably sensory. On the eye tubercle, between the lateral eyes, of *Endeis spinosa* there is a patch of thin cuticle which protrudes from the surface but no function has yet been attributed to it (Dohrn, 1881). On the ovigerous legs of many species such as members of the genera *Colossendeis* and *Nymphon* there are spines which have a nerve supply and branched or spatulate distal ends (Figs. 7i,k and 8c). Although these structures may have a mechanical function, such as helping to grip the eggs or cleaning the surface of the body, they may also have a tactile function. This may also apply to spines with articulated bases which occur on various parts of the body in many species.

Protruding from pores over most parts of the body, and often arranged in rows along the longitudinal axis of the proboscis in some species, e.g. *Endeis spinosa,* are small bifid hairs (Fig. 17b). In some species such as *Achelia echinata* the branches are short and subdivided with other short side branches (Fig. 17b). The form of branching is often species specific. They may be chemoreceptors or tactile sense organs and perhaps some of these hairs enable the pycnogonid to detect currents over the body surface, particularly in those examples where they are long and thin. There is no evidence concerning the means of detection of *pressure* change (see page 35). In other species such as *Nymphon gracile* or *Callipallene brevirostre* the bifid hairs are long and thin (Fig. 17b).

On parts of the body in species such as *Pycnogonum hancocki*

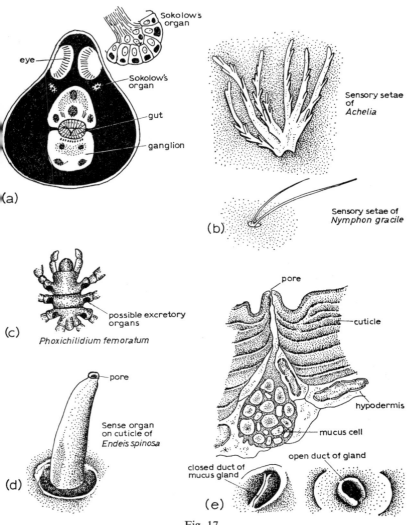

Fig. 17

(a) Section through the region of the brain showing the position of Sokolow's organ. (Drawn from photographs by Hanstrom, 1965)
(b) Two types of sensory hairs which are scattered over the surface of the body in some species, the thicker branched type characteristic of members of the genus *Achelia* and the long thin types present in the genus *Nymphon*.
(c) The position of possible excretory organs on the lateral protuberances of *Phoxichilidium femoratum*.
(d) Seta with a terminal pore from the palp of *Endeis spinosa*. This is possibly a chemoreceptor.
(e) Section through the cuticle showing mucus cells and duct and inset a surface view of the duct opening in closed and open position.

and members of the genus *Achelia,* spines are present with a small terminal pore (Fig. 17d). These are particularly abundant on the palps of members of the genus *Achelia.* Although there is no direct evidence regarding their function, similar structures in other arthropods have had a chemosensory function attributed to them (Slifer, 1969).

REPRODUCTIVE SYSTEM

In pycnogonids, except for one record of a hermaphrodite species, *Ascorhynchus corderoi* (Marcus, 1952) (Fig. 18i), the sexes are separate. In each sex the reproductive organs consist of a pair of ovaries or testes lying above the gut on either side of the heart (Fig. 18b,c). In the adult the two parts are fused posteriorly at the base of the abdomen to form a U-shaped structure. In *Phoxichilidium*, fusion of the right and left components extends forwards so that the ovary forms a thin, broad plate. Diverticula pass into the ambulatory legs, those of the testis reaching to the third joint of the legs and those of the ovaries to the fourth or sometimes further. In most species, the ova ripen within the lateral diverticulae, usually in the femurs but occasionally in one or more coxae as well. In *Pycnogonum littorale,* the ova also ripen in the trunk part of the ovary (Fig. 18c). In the legs, the ovarian tissue is pressed closely to the dorsal and dorso-lateral wall of the gut caeca (Fig. 18a), and the gut cells may

---

Fig. 18

Diagrams showing the extent of the reproductive system in males and females, the structure of the genital openings as seen on the stereoscan electron microscope, the sperm and internal arrangement of the eggs above the gut.
(a) Longitudinal section of part of the femur of *Nymphon gracile* showing the ovary containing eggs dorsal to the gut.
(b) Diagram showing the shape of the ovary in *Nymphon gracile*.
(c) Diagram showing the shape of the ovary in *Pycnogonum littorale*.
(d) Diagram showing the shape of the testis in *Colossendeis proboscidea*. (Redrawn from Hoek, 1881)
(e) Female genital opening of *Achelia simplex*.
(f) Female genital opening on second coxa of the legs of *Nymphon gracile*.
(g) The coxal projection on the second coxae of the third and fourth pairs of legs in the male of *Achelia simplex*. The genital opening at the tip.
(h) Male genital opening on the second coxa of *Nymphon gracile*.
(i) Section through the second coxa of the fourth leg of the adult hermaphrodite pycnogonid *Ascorhynchus corderoi*. (Redrawn from Marcus, 1952)
(j) Sperm of *Achelia echinata*.
(k) Sperm of *Phoxichilidium femoratum*.

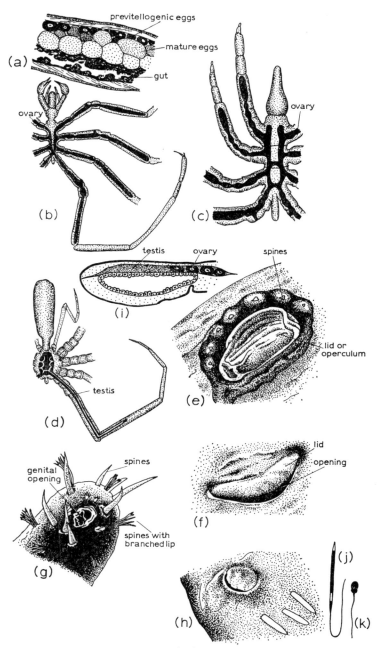

Fig. 18

have a role in preparation of yolk materials. Whether germ cells move from the trunk to the diverticula of the ambulatory appendages in species where all vitellogenesis occurs within these appendages, is not known.

Each genital diverticulum communicates with the exterior through a duct, with a valvular opening (Fig. 18b,c,e,f,g,h), situated on all the second coxae or some of them (see page 64). In most species, the pores have a lunulate-shaped aperture. The closing device is in the form of a lid or operculum. It is possible that the muscular contraction of the walls of the diverticula force their products against the operculum and open it. The elasticity of the cuticle and the presence of muscles are probably sufficient to close the operculum when the eggs have passed out (King and Jarvis, 1970). In some species, such as *Achelia echinata,* the genital opening lacks the typical operculum and is situated at the tip of a protuberance (Fig. 18g).

A detailed description of egg development within the ovary has been made for only two species, *Nymphon gracile* (King and Jarvis, 1970) and *Pycnogonum littorale* (Jarvis and King, 1972a). The process of egg development is basically similar in both these species. For convenience the stages of developing oocytes may be divided into the previtellogenic and vitellogenic phases. The vitellogenic phase usually occurs when the oocytes are in a ventral position in the ovaries and, in most species, in the ovarian caeca within the femur and sometimes the coxae of the ambulatory appendages, and thus in a position proximal to the gut. The size of the eggs varies between species depending upon how much yolk is present. The eggs are small in *Phoxichilidium* sp. and *Tanystylum* sp. having a diameter of only 0·05mm, larger in *Callipallene* sp. with a diameter of 0·25mm and considerably larger in *Nymphon* sp. with a diameter of 0·5 or 0·7mm (Sauchez, 1959). This is, however, relative to the size of the body. Hence the eggs of *Callipallene* sp. are proportionately much larger than the others.

The previtellogenic oocyte (Fig. 19) is surrounded by a thin basement membrane consisting of amorphous material. The membrane is relatively smooth without microvilli, which are usually characteristic of the periphery of arthropod eggs, such as those of *Limulus polyphemus* at this stage (Dumont and Anderson, 1967). The cytoplasm is densely packed with ribosomes which become progressively more scattered as oocyte growth proceeds. The nucleus has a number of adielectronic aggregations of material in its periphery and some of this material occurs on the outer side of the envelope. This suggests that it passes out of the nucleus into the cytoplasm. As growth continues, a single, centrally-situated large adielectronic inclusion appears in the nucleoplasm.

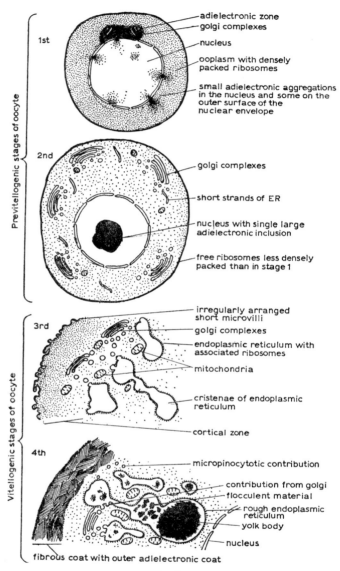

Fig. 19

Diagrammatic representation of the cellular processes involved during egg development. Previtellogenic and vitellogenic phases are shown and the process is arbitrarily divided into four stages.

Some annulate lamellae are present in the same zone as the golgi complexes but since they were first observed in *Arbacia* oocytes, they have been given a variety of names. They have been more frequently observed in the developmental stages of both male and female germ cells, in embryonic cells and neoplastic cells, though they do occur in some normal somatic cells (Kessel, 1968). In most instances, the annulate lamellae arise from the nuclear envelope and evidence suggests that they contribute to the formation of smooth endoplasmic reticulum. Their appearance in the previtellogenic oocytes of *Nymphon gracile* is followed by a rapid increase in the amount of endoplasmic reticulum in the ooplasm and perhaps they, associated with the golgi complexes, are responsible for the production of the endoplasmic reticulum network which quickly fills the cytoplasm.

In the early previtellogenic oocyte there is a zone near the nucleus which is more adielectronic than the rest of the cytoplasm. This region contains most of the mitochondria of the oocyte and the golgi complexes first appear there, so that this area resembles the yolk nuclei or 'Balbiani bodies' described in a variety of animal oocytes. It is now believed that the term 'Balbiani body' denotes not a definite structure but an aggregation of varying combinations of organelles such as mitochondria, whorls of endoplasmic reticulum, golgi complexes, ribosomes and annulate lamellae, depending upon the species involved. A mixture of ribosomes and mitochondria occurs in this region in the annelid *Spirorbis borealis* (King, Bailey and Babbage, 1969) and in spiders and *Artemia,* the zone consists of endoplasmic reticulum, mitochondria, golgi complexes, ribosomes and multi-vesicular bodies (Raven, 1961; Nørrevang, 1968). This is a more complex combination than the condition in *N. gracile* which resembles that in *S. borealis*. These organelles combined can provide templates and energy for the production of structural and storage materials.

The formation of yolk in oocytes has been studied in a wide range of animals. In some animals, a large amount of it is manufactured outside the ovary and is carried to the developing oocytes by the blood or haemolymph and then passed into the eggs either by a specialised type of pinocytosis at the oocyte plasma membrane, or having entered auxiliary cells is passed through cytoplasmic connections from these cells into the egg. This type of yolk formation has been described in fish, amphibia, birds, early stages of mammals, polychaetes, insects and to some extent in *Limulus polyphemus*. In some other animals the material from which the yolk is formed presumably diffuses into the oocyte in a low molecular form, since no pinocytotic activity or cytoplasmic connections have been observed. This type of vitellogenesis has been described in *Priapulus,*

Anodonta, S. borealis and in the early stages of vitellogenesis in *L. polyphemus* and *L. emarginata*. This is probably the primitive method of yolk formation in the egg and all intermediate stages exist from this to almost completely external formation of yolk materials as in insects. In *N. gracile*, in common with *L. polyphemus*, the yolk precursors enter the egg in low molecular form though later in vitellogenesis a very limited amount of pinocytosis occurs. This suggests a fairly primitive arthropodan type of vitellogenesis and as in other animals with the type of vitellogenesis the time taken to produce an egg is considerably longer than in the more advanced types (King and Jarvis, 1970). In *N. gracile* the eggs develop over winter (Fig. 21), egg development starting in young individuals in July. A few eggs have reached the vitellogenic phase by November and the majority do so after March the following year. This is presumably correlated with the spring abundance of food materials (King and Jarvis, 1970).

Within the eggs of various animals, yolk formation has been attributed to a number of different organelles. In some the yolk is formed within mitochondria as described in the eggs of *Planorbis, Limnea* and *Mytilus* (Nørrevang, 1968). In others the yolk is formed freely in the cytoplasm as in *Priapulus* (Nørrevang, 1965), *Patella coerulea* (Nørrevang, 1968) and spiders, or it may be formed in intimate contact with vesicles of the endoplasmic reticulum. In others it is formed within pre-existing vacuoles which is the method which occurs in insects and to some extent in *Lebistes* (Droller and Roth, 1966). Fourthly, the formation of the yolk occurs within cisternae of the endoplasmic reticulum as in crayfish *Astacus* (Beams and Kessel, 1963), a guppy *Lebistes* (Droller and Roth, 1966) and *Spirorbis borealis* (King, Bailey and Babbage, 1969), though the details of organelle involvement differ in all these examples. In the crayfish, large electron-dense granules form within cisternae and move into the agranular parts of the endoplasmic reticulum where they coalesce (Beams and Kessel, 1963).

The process starts near the oocyte nucleus and proceeds towards the periphery, where the first fully-formed yolk bodies appear. In *Lebistes* (Droller and Roth, 1966), flocculent material appears within the cisternae of the endoplasmic reticulum prior to the onset of the pinocytotic activity and this material moves into regions of smooth endoplasmic reticulum where it coalesces to form the yolk bodies. A similar process has been described in the eggs of *Spirorbis borealis* though in both instances a definite contribution to the forming yolk is made by the golgi complexes but, unlike the process in *Astacus*, the first fully-formed yolk bodies appear near the oocyte nucleus. A similar process occurs in the eggs of *N. gracile* and *P. littorale* in which the first discrete yolk bodies, initially containing polysaccharides and proteins and later lipids, first appear in a perinuclear zone (King and Jarvis, 1970; Jarvis and King, 1972a). When the first vesicles are formed associated with endoplasmic reticulum, they contain some scattered small aggregations of medium electron-dense flocculent material. As egg development proceeds, the network of endoplasmic reticulum contains large amounts of this material and within the cisternae larger, centrally-placed masses occur, presumably as a result of coalescence of the initially scattered smaller aggregations. During development, these masses increase in size and the amount of scattered flocculent material is reduced (King and Jarvis, 1970) (Fig. 19).

At the onset of vitellogenesis the golgi complexes arrive in their semiperipheral position and numerous vesicles appear associated with them and probably budded from them. These vesicles become associated with the endoplasmic reticulum and fusion occurs until a complex branching system of tubular and vesicular elements of the endoplasmic reticulum results. Some distance from the periphery of the oocyte, the cisternae and tubes of the endoplasmic reticulum become coated with ribosomes with a corresponding decrease in the free ribosomal content of the cytoplasm in this area. Mitochondria are at this stage in vitellogenesis more numerous in the oocyte and are more evenly distributed throughout the cytoplasm (King and Jarvis, 1970, 1972).

In some electron micrograph profiles, other small vesicles containing flocculent material are visible fusing with cisternae (Fig. 19). This suggests that some other organelle is contributing to the formation of the yolk. In the earlier stages of vitellogenesis, this agent is probably the still active golgi complexes, but later, limited micropinocytosis occurs at the oocyte periphery, which seems to contribute to the formation of the yolk by fusion of the pinocytotic vesicles with the endoplasmic reticulum cisternae, containing the nuclei of flocculent material (King and Jarvis, 1970).

Although there is little evidence of pinocytotic activity in the early stages of vitellogenesis in *Nymphon* a number of small pits appear later and, after closure, move through the cortical region of the egg. This activity persists until the end of vitellogenesis. Coinciding with vitellogenesis, a layer of material is formed between the basement membrane and the surface of the oocyte. This consists of numerous fine fibres, which are produced from the surface of the oocyte and consist of a protein/carbohydrate complex with some lipid associated, and also containing acid mucopolysaccharide. The apparent close contact between these layers of adjoining oocytes suggests an adhesive function for this layer and perhaps it helps the eggs to stick together so that they can be picked up by the male when they leave the genital pore on the second coxa of the female. The relative importance of this layer and the products of the cement glands on the femur of the male in some pycnogonids has not been determined (see page 43). Although the process has not been observed, it seems likely that the eggs are forced out of the pore by peristalsis of the muscles surrounding the ovary and the layer round the oocytes may act as a lubricant in order to protect the eggs from damage during their release. The eggs of most pycnogonids have this fibrous or hyaline layer around them which solidifies rapidly in contact with water (King and Jarvis, 1970).

4

LIFE CYCLE

Fecundation in pycnogonids has been described in several different genera, though the fertilisation of the eggs has not been observed. Mating has been described in *Anoplodactylus lentus,* which reproduces towards the end of August on the coast of North America (Cole, 1901). The male climbs on to the back of the female, then crawls over her head to the ventral surface so that the ventral surfaces are opposed to one another and the male and female are orientated head to abdomen. They then cling together with their ventral surfaces and genital apertures opposed. The hooked ovigerous legs of the male then fasten on to the extruding egg-masses and tear them away and by a rotational movement form them into a ball. Fertilisation is believed to occur externally as the eggs are released by the female but before the hyaline coat is affected by the sea-water. A similar process has been described in *Phoxichilidium femoratum* (Loman, 1907) and *Endeis laevis* (Hoek, 1881; Prell, 1910). A different process occurs in *Pycnogonum littorale.* In this species the male rests for a long period on the back of the female with their genital orifices touching. These orifices are situated on the ventral surface of the second coxa of the last pair of legs in the male and on the dorsal aspect of the same leg and segment in the female. When the eggs are released they form a single extensive ball in this species with both ovigerous legs embedded in it. Most other species (Fig. 20b) have a number of balls of eggs on each ovigerous leg of the males during the breeding season. In *Phoxichilidium femoratum*, each ball of eggs, which the male carries, represents the entire brood of one female and some males have been observed carrying fourteen balls of eggs. Females of *Endeis spinosa* release a mass of eggs at each mating which represents the contents of a single femur; *Nymphon gracile* release the contents of two femurs and *Callipallene,* which

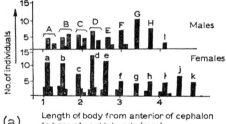

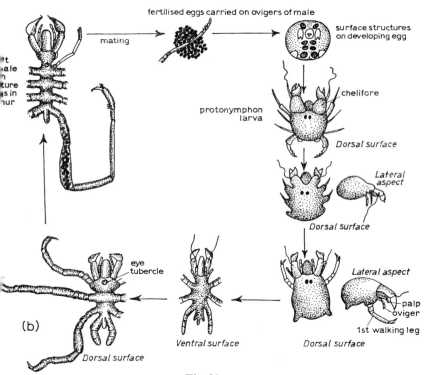

Fig. 20
(a) Trunk lengths of males and females of *Pycnogonum littorale*: A–I, proposed intermoult size groups for males; a–k, proposed intermoult size groups for females.
(b) Diagram of the life cycle of a typical pycnogonid. Eggs released by the female and after fertilisation carried on the ovigers of the male. Hatch into a protonymphon larva. Segments added at the posterior end as growth takes place until adult stage reached. (Drawings not to same scale)

produces only one or two eggs in each femur, releases the contents of all femurs at the same time. The balls of eggs carried on each oviger by a single male of Nymphon gracile have obviously been obtained at different times because there is usually considerable difference in the stages of development of the eggs in adjoining balls on one oviger, and some males have a ball of newly released eggs on an oviger alongside a ball of protonymphon larvae.

Littoral species tend to have a seasonal release of eggs as shown by *N. gracile* but off-shore species frequently have some eggs released at all times of the year (Fig. 21). Specimens of *Achelia echinata*

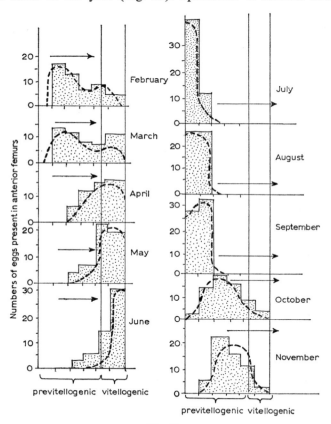

Fig. 21

Histograms showing the size and number of eggs present in the anterior femurs of females of *Nymphon gracile* during months on the shore. The sizes of the eggs, measured as diameters, are divided into previtellogenic and vitellogenic phases.

Life cycle

which occur on the shore have a period of egg release similar to that of *N. gracile* but members of this species living off-shore have some release of eggs throughout the year.

The method of segmentation to form a blastula, which occurs after entry of the sperm into the egg, follows the pattern characteristic of other arthropods and depends upon the amount of yolk present in the egg. Members of the genera *Phoxichilidium, Anoplodactylus, Ammothea* and *Pycnogonum* are typical of pycnogonids having small eggs containing a moderate amount of yolk. In the eggs of these pycnogonids the division of the egg is complete and cleavage results in the formation of equal-sized blastomeres (Fig. 22). Division continues until a ball of cells is produced, which has a very restricted blastocoele at its centre (Fig. 22). In the eggs of members of the genera *Nymphon* and *Chaetonymphon*, which are larger and have more yolk than the others described, the division is total but unequal. Cleavage continues until a ball of cells is formed consisting of small micromeres and large macromeres (Fig. 22). A third type of division has been recorded in eggs which are very rich in yolk. This type is characteristic of species belonging to the genus *Callipallene*. In these eggs the initial divisions are complete but later ones are only partial and eventually a periblastula is formed. This method of division is sometimes referred to as superficial segmentation and at the stage of segmentation, when eight cells have been formed, the upper plaque of four does not correspond in orientation with the four beneath them. This arrangement has led to the suggestion that during cleavage of at least some pycnogonid eggs a rudiment of spiral cleavage persists (Sanchez, 1959). This does not normally occur in the arthropods although it does in annelids, and is interesting since the arthropods are thought to have arisen from the ancestral annelids.

The formation of the germ-layers in eggs which have little yolk, such as those of the genera *Pycnogonum, Phoxichilidium* and *Anoplodactylus*, occurs by a process of inward growth of cells into the blastocoele. A blastomere, which is initially clearly visible on the outside of the ball of cells or blastula, grows into the blastocoele and by later division forms the endoderm (Fig. 22). At the same time, other cells of the blastoderm situated near to the initial position of this endoderm rudiment sink inwards and form the rudiment of the mesoderm (Fig. 22). Thus gastrulation takes place by an inward growth of cells. The position of this in-tucking or epibody is interesting as it occurs in a position which corresponds with the future dorsal side of the adult animal (Dawydoff, 1928). This is thought to be a primitive condition. When this region of in-tucking, sometimes referred to as a blastopore, is closed, the gastrula consists

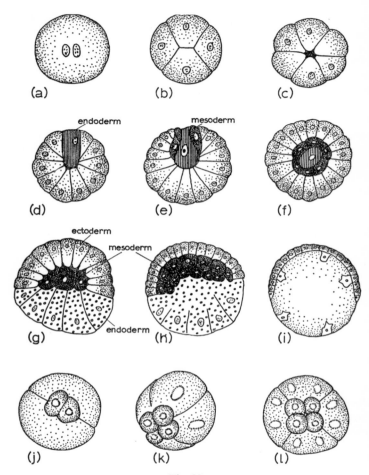

Fig. 22

Formation of the blastula and the gastrula in specimens with little yolk in the eggs and those with a lot of yolk. The examples given are *Pycnogonum littorale* (a–f) and *Nymphon* with a moderate amount (g–h), and *Pallene* (i) and *Callipallene* (j–l) with abundant yolk.

of a group of centrally situated endodermal cells surrounded by a mass of mesodermal cells. The endodermal cells then fuse to form a syncytium (Dawydoff, 1928; Sanchez, 1959).

In those species which have eggs containing abundant yolk, and in which there is unequal segmentation during the formation of a blastula, gastrulation occurs by a process of overgrowth or epiboly. By this process the larger cells or macromeres are gradually covered by an overgrowth consisting of the smaller cells or micromeres. In eggs with this type of gastrulation, the mesoderm is formed at the edges of the blastopore and it is at this stage of development, when the lip of the blastopore is in an equatorial position, that some macromeres become isolated and place themselves under the dome of the micromeres. The cells then form the rudiment of the mesoderm initially consisting of about four cells and the endodermal blastomeres then fuse and form a syncytium which is rich in yolk. Inside this syncytium are a number of active centres which constitute vitellophages (Dawydoff, 1928).

No satisfactory description has been made of gastrulation in those pycnogonid eggs which have a superficial type of division and in which the embryo is formed only on part of the egg. This type of development differs from that in the other pycnogonid eggs which have less yolk, in which the yolk is enclosed by the developing embryo (Fig. 22).

Following gastrulation the ectoderm forms five separate pairs of thickenings which are the same number as the initial ganglia of the nervous system. The first of these pairs forms above the stomodeum, on the dorsal side of the embryo, and immediately fuses to form the protocerebrum (Figs. 23 and 24). The other pairs of ganglia are situated on the ventral side, and the first of these corresponds to the first pair of appendages, the chelifores. Although this pair is initially post-oral in position it moves to a dorsal position and fuses with the protocerebrum in most species, though in some it remains in a position lateral to the stomodeum. The ganglia of the second and third post-oral segments correspond to the second and third appendages and soon fuse to form a ganglionic mass. This is the suboesophageal ganglion. The fourth pair of ganglia appears later than the others and remains unutilised until the fourth pair of appendages develops (Dawydoff, 1928) (Fig. 24).

Mesoderm. The question of mesoderm in the pycnogonids is still unresolved. Originally, the formation of mesodermal bands was described in *Pallene* and the segmentation was said to be shown by the formation of coelomic pouches penetrating into the limb buds, but all the more recent observations categorically deny the formation of a coelom. One thing seems certain, the greater part of the

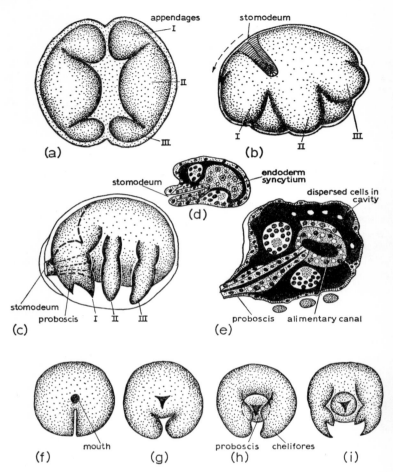

Fig. 23

Series of diagrams showing the formation of the larval appendages (I–III), the initial dorsal position of the stomodeum and its subsequent movements, and stages in the formation of the alimentary canal from a syncytium of endoderm. (a–i after Helfer and Schlottke, 1935)

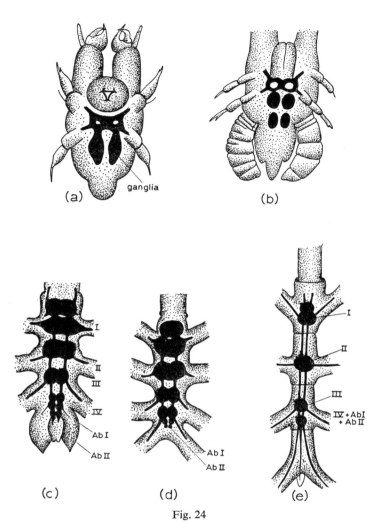

Fig. 24

(a–e) A series of diagrams showing the stages in the development of the nervous system of *Endeis spinosus* from larva to adult. (After Helfer and Schlottke, 1935)

mesoderm is concentrated in the appendages and it is almost entirely used up in the formation of the muscles. The small amount of mesoderm which is not used for muscle formation remains as isolated cells dispersed in the body cavity of the embryo. These cells later give rise to the walls of the dorsal blood vessel or heart and to the blood corpuscles. Amongst the mesodermal cells, there are some which, much later in development, form the reproductive system (Dawydoff, 1928).

Alimentary canal. The stomodeum appears early in development but the proctodeum does not appear until a short time before the eclosion of the larva. With regard to the midgut, in pycnogonids with small eggs, the formation is relatively simple but is more complicated in eggs which are rich in yolk. One characteristic common to all pycnogonids is that the endoderm forms a syncytical mass (Fig. 23). This tendency, which is very pronounced in those forms with large eggs, also occurs in those in which the endoderm retains its primitive cellular character for a long time. This syncytium transforms itself into a thick-walled sac, which connects with the tubular-shaped stomodeum. The combined units, forming a gastral sac, remain in this condition for a long time (Dawydoff, 1928).

In eggs which are rich in yolk the midgut is formed from endodermal elements dispersed in the interior of the mass of yolk, where they are mixed with vitellophages. These endodermal cells finish by forming an epithelial layer at the periphery of the yolk.

During the post-embryonic period, direct development has been observed in some species of *Pallene* and *Nymphon*. In *Nymphon brevicaudatum* the embryos leave the chorions when they possess the full complement of adult appendages. In other pycnogonids, however, the animal which ecloses is much less advanced than this stage and needs to undergo further change to attain the adult form. This early stage is called a protonymphon (Fig. 25). Exteriorly, the protonymphon larva characteristically has three pairs of appendages and superficially resembles a nauplius larva of the Crustacea. The appendages represent the chelifores, palps and ovigers of the adult and all consists of three segments.

The chelifores which are the only ones having a dorsal position terminate with a pair of pincers, although the other appendages only have a single terminal claw. One finger of the pincer is fixed and the other moves to appose it by the action of muscles attached at its base. All the appendages have a long spine attached to their proximal segment.

The protonymphon is square to ovoid in shape. It has a ventral proboscis and has eyes on the dorsal surface. The larva is segmented internally although there is no trace of this externally. It consists of a

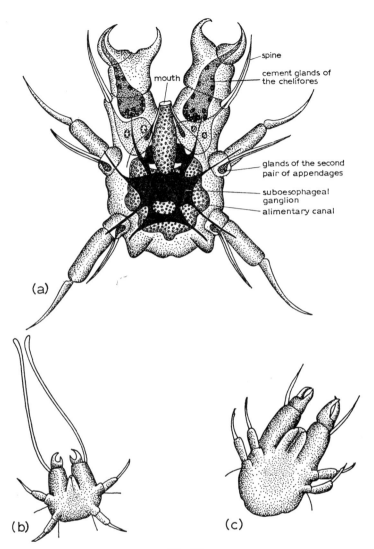

Fig. 25
(a) Structure of the Protonymphon larva of *Achelia echinata*. External views of the ventral surfaces of the larvae of *Pycnogonum littorale* (b) (after Helfer and Schlottke, 1935 (c) (after Dogiel, 1911)

pre-oral region and four post-oral segments, of which the anterior three bear the appendages, which are well developed, and the posterior one only has two small papillae, which are the buds of the fourth pair of appendages. In addition to these, there is an unsegmented abdominal section (Dawydoff, 1928). The larvae of some pycnogonids have a number of modifications compared with the more generalised form. There are temporary larval organs in *Pycnogonum* sp. in which the spines of the chelifores are transformed into long filaments by which the larva is able to attach itself to the colony of hydroids on which it is believed to undergo metamorphosis (Fig. 25b). When this has occurred, the flagellum disappears. In *Phoxichilidium* sp. and *Anoplodactylus* sp., in which the chelifores have lost the cement glands, the second and third pairs of appendages are modified to form long filaments, often five times as long as the limb itself (Fig. 25b).

Pycnogonids characteristically have an anamorphic type of development. In this type of development the larvae hatch with few segments or metameres and the post-embryonal appearance of additional metameres increases the number until the final number characteristic of the adult is reached. Within the arthropods this type of development occurs amongst the trilobites, crustaceans, many myriapods and the proturans. Upon eclosion the proturans and myriapods have, if not quite the full adult numbers, at least all the functional types of metameres and appendages. In the Crustacea the nauplius hatches with only three pairs of appendages and these have to perform the functions which are later carried out by more than ten pairs in the adult. The form of the protonymphon larva in the pycnogonids is similar to that of the nauplius larva except that little is expected of it except to adhere to the ovigers of the male.

A larva which forms from the type of egg which has little yolk soon leaves the ovigers of the male to pursue a free life. On the other hand there is a tendency for those which are formed from eggs which have plenty of yolk to remain fixed for a longer period to the ovigers of the male. In some species attachment to the ovigers is maintained until metamorphosis is complete.

The external phenomena associated with metamorphosis vary in detail between species but they possess a common basic pattern. There is always a series of moults after each of which a new pair of appendages form behind the pre-existing ones and the larval appendages are gradually reduced as the adult ones appear, though views differ concerning the relationship between the larval and adult appendages (Fig. 20b). Some workers believe that the former are converted to the latter, whilst others believe that the adult appendages are new structures unrelated to those of the larva (Sanchez, 1959).

Life cycle

The internal phenomena associated with metamorphosis consist of the formation of the dorsal blood vessel and of the reproductive system. In addition to this there is the organisation of new segments, each of which requires muscles and ganglia. The new ganglia are produced by proliferation of the ectoderm (Dawydoff, 1928) (Fig. 24). The origin of new mesodermal elements is difficult to locate but it is probable that new mesoderm is formed independently of the old. The primitive mesoderm, as already stated, is almost entirely used up in the formation of the larval organs, but some free elements remain and these produce the heart and gonads. The latter appear as segmentally arranged buds which later fuse to form the adult structure (Dawydoff, 1928).

5

FEEDING

The food preferences and possible range of diets for species of pycnogonids are very poorly known. Most of the evidence is based solely on observations of associations between pycnogonids and other animals or plants they have been collected with. They are most frequently found with polyzoans or hydroids. The types of association between pycnogonids and polyzoans may fall into one of three categories (Wyer and King, in prep.):

(1) The polyzoans merely provide a firm substrate to which the pycnogonids can attach themselves and on which they safely move about, an association exemplified by *Pycnogonum littorale* and *Endeis spinosa* on the polyzoan *Flustra foliacea*.

(2) The pycnogonids feed on epiphytes, epizoites or debris on the surface of the polyzoan. Examples of this are *Achelia longipes* feeding on shoots of red algae growing on the polyzoan *Flustrellidra* which was itself growing on the surface of the alga *Gigartina,* and *Nymphon rubrum* feeding on *Laomedea angulata* which was growing on *F. foliacea*.

Fig. 26
- (a) The proboscis and non-chelate chelifores of *Phoxichilidium femoratum*.
- (b) *Pycnogonum littorale* with its proboscis embedded in an anemone.
- (c-g) A range of mouth shapes correlated with the type of food eaten.
- (c) *Achelia echinata* which feeds on polyzoans such as *Flustra*.
- (d) Horny lip of *Pycnogonum littorale*.
- (e) Mouth of *Pycnogonum littorale*.
- (f) Mouth of *Anoplodactylus angulatus*.
- (g) Mouth of *Endeis spinosa* which is a browser.

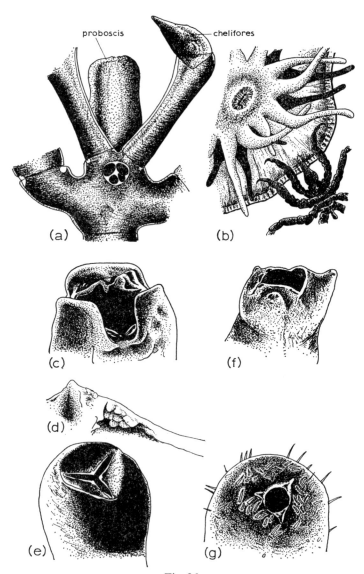

Fig. 26

(3) Pycnogonids actually feeding on polyzoans, e.g. *Achelia echinata* on *F. foliacea, Nymphon gracile* on *Bowerbankia, P. littorale* on rotting edges of *F. foliacea* (Wyer and King, in prep.) and *Australodecus glaciale* on *Cellarinella* (Fry, 1965). *A. glaciale* attacks the polyzoan by inserting its proboscis through the frontal pore.

There are numerous instances of pycnogonids observed feeding on hydroids. *Phoxichilidium femoratum* has been seen feeding on *Tubularia larynx, Bougainvillea, Endendrium, Clava, Coryne* and *Syncoryne*; *Anoplodactylus lentus* on *Endendrium ramosum*; *Nymphon gracile* on *Actinia equina*; *Nymphon leptocheles* on *Campanularia* and *Dynamena pumila*; *Pycnogonum littorale* on *Tealia crassicornis, Lucernaria, Cucumaria frondosa, Metridium dianthus* and *Metridium senile*; and *Pycnogonum stearnsi* on *Anthopleura xanthogrammica* and *Bunodactis elegantissima*.

The method by which food is passed into the mouth varies, depending upon the armature around the mouth, shape of the proboscis and whether the species has chelifores or palps. When *N. gracile* feeds on the hydroid *D. pumila* it clings to the colony by means of its long, slender legs and removes pieces of the hydroid by using the right and left chelifores alternatively (Fig. 27a). The palps are used to locate the position of the hydroids then pieces are cut from the distal ends of the hydroids by the chelifores, partly macerated by a rapid scissor-like action of the chelae and then held against the mouth so that they can be seized by the jaws and drawn into the mouth. A similar method is used when this species of pycnogonid feeds on the polyzoan *Bowerbankia* or small sedentary polychaetes. A second method is used, however, when the food substrate is large in relation to the pycnogonid such as *Actinia equina*. This involves grasping the sides of the actinian and rasping away pieces of its body wall by the action of the jaws of *Nymphon gracile* or the suction and rasping jaws of *Pycnogonum littorale* on *Tealia* (Figs. 26b,d,e).

Fig. 27

Illustrations of the feeding methods of pycnogonids which use the chelifores to bring the food to the proboscis.
(a) *Nymphon gracile* holding hydroids with its chelifores. Support to the food is also given by the palps as it is brought to the mouth.
(b) *Endeis spinosa*.
(c) *Anoplodactylus angulatus*.
(d) *Achelia echinata* waiting above the polyzoan *Flustra* for the operculum to open. Legs omitted from one side of pycnogonid for clarity.
(e) Proboscis of *Achelia echinata* thrust into operculum of *Flustra*.
(f) Body of *A. echinata* moved forwards so that proboscis levers open the operculum.

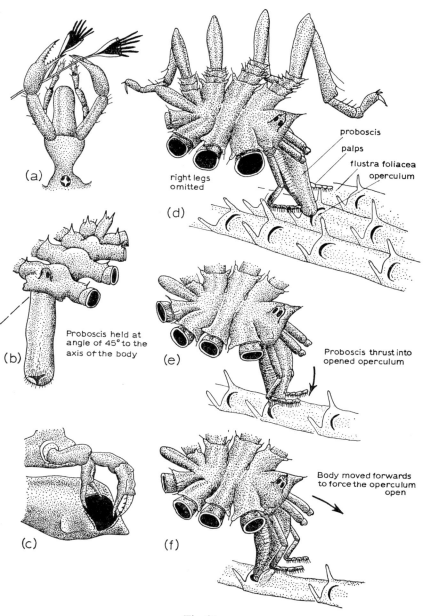

Fig. 27

Nymphon rubrum feeding on the hydroid *Laomedea angulata* has been studied. One chelifore is used to hold the stem while the other tears a hole in it (Fig. 27a). When this has been done the proboscis is placed against the hole and the tissues within sucked out. When feeding is completed in both *N. gracile* and *N. rubrum* the chelifores and proboscis are cleaned by grooming with the ovigers.

Whilst the proboscis of *N. gracile* is held against the food, material containing acid phosphatase is passed forwards from the midgut and down the proboscis until it comes into contact with the food. This movement is caused by pulsations at a rate of 160 per minute in the region of the midgut in the trunk but not in the caeca. The absence of pulsations in the caeca is probably correlated with the presence of enlarged muscles around the gut near to the limb bases which, when they contract, may act as valves isolating the lumen of the caeca from that of the midgut.

Gradually the food becomes lighter in colour and looser in texture which suggests that enzyme activity has occurred. It is then taken in at the tip of the proboscis, the jaws periodically opening to suck in the food, which has by this time become a mass of loose particles.

Once the food has been taken in, it is moved around within the proboscis by the action of the dilator and contractor muscles (Fig. 5). The food is slowly moved towards the spines which form a sieve at the base of the proboscis. It then passes through the oesophagus to the midgut by peristaltic movements. The consistency of the food at this point is still finely particulate and it passes into the caeca within the ambulatory legs from the base to the tip, with slight backflow between pulsations (Fig. 14a). After this general distribution, the food undergoes the rest of the digestive process.

Polyzoans have been found with holes in their frontal walls and it is possible that these were produced by digestive enzymes of pycnogonids. When *Phoxichilidium femoratum* or *Anoplodactylus petiolatus* feed on *Dynamena pumila* they seize the stems of the hydroid using the apposition of the propodal sole spines and the terminal claws of the legs (Fig. 26a). Pieces are torn from the hydranth by the chelifores and transferred to the mouth where they are ingested. Thus in both the Nymphonidae and the Phoxichiliidae the food is gathered primarily by the chelifores which are of the right length and have their chelae positioned usually above the proboscis so that they can pass the food to the mouth (Figs. 26a,f and 27a,e). The proboscis in members of these families is usually cylindrical and the extrinsic musculature is not adapted for fine control.

The Ammotheidae lack functional chelifores in the adults in which they are represented solely by a rudimentary scape without func-

Feeding

tional chelae (Fig. 40). The proboscis and associated extrinsic musculature is modified from the simpler condition in the Nymphonidae and Phoxichilidiidae.

When *Achelia echinata* feeds on *D. pumila* the hydrotheca is located by the palps and the tapering proboscis is inserted and parts of the tentacles grasped by the jaws. The proboscis is then slightly withdrawn until the tentacle is snapped off when injection follows. When *A. echinata* feeds on the polyzoan *F. foliacea* the palps are used to locate the food (Fig. 27d). The position of the spines of the polyzoan zooecium are used to locate the operculum so that the pycnogonid may wait above it and when it opens the proboscis tip inserted (Fig. 27d). This is done at an angle to get under the lid of the operculum then when entry has been achieved the angle of the proboscis is changed so that the operculum is prised open (Figs. 27e,f). A related species, *Achelia longipes,* also has a mobile proboscis and fine jaws but in this pycnogonid the proboscis is used to graze small shoots of red algae which grow on the surface of polyzoans (Fig. 26c). The surface of the polyzoan is explored with the palp until some soft shoot of alga is located when the proboscis is lowered so that feeding can take place. *Australodecus glaciale*, which has a long thin proboscis, feeds on the polyzoan *Cellarinella foveolata* by inserting its proboscis through the frontal pore (Fry, 1965).

The Endeidae are without palps or chelifores in the adult stage so that all food location and gathering must be performed by the proboscis (Fig. 27b). The mouth in this family is surrounded by numbers of spines or setae which presumably have a sensory function (Fig. 26g). These carry out the function of the palps in the Nymphonidae and Ammotheidae. The proboscis of some species of the Phoxichilidiidae has a number of small protuberances which may have a sensory function but these are fewer and not so obviously suited for a sensory function as the ones present in the Endeidae.

In *E. laevis*, which feeds on the detritus frequently occurring at the bases of colonies of hydroids such as *Antennularia*, the proboscis is normally held pointing towards the substratum at an angle of 45° to the longitudinal axis of the body (Fig. 27b). It is pushed into recesses in the hydroid and moved slightly from side to side. This breaks up and loosens the detritus by the spines at the tip of the proboscis which facilitates the uptake. Some areas of debris are ingested in preference to others which supports the idea of sensory receptors at the proboscis tip.

The Pycnogonidae are without both chelifores and palps (Fig. 38a). Most species feed on hosts which are large compared with themselves. The first pair of legs are used to grasp the prey, the tip of the proboscis is pressed against it and then the second, third and fourth pairs of legs are used to push against the prey (Fig. 26b). This pushing combined with the rasping action of the jaws forces the tip of the proboscis into the tissues of the prey. The proboscis of *Pycnogonum littorale* has a wide base which tapers distally then dilates again slightly (Figs. 26e and 38a). The narrowed part is covered by smooth cuticle and when the body of the prey has been penetrated by it the tissues close onto it tightly so that the sucking action caused by a slight dilation of the pharynx is aided by the pressures of the prey's tissues. In some species with a large mouth, e.g. *Pycnogonum hancocki*, the cuticle has a reticulation of thickenings in the cuticle presumably to give it greater strength to enable greater stress to be put on it by pushing without buckling occurring (Fig. 4e). Little is known concerning the mode of feeding in either the Colossendeidae (Fig. 37) or the Tanystylidae.

Food preferences frequently depend upon the type of food on which the pycnogonid has previously fed. This is usually taken in preference to food not previously sampled but which may be preferred by individuals of the same species of pycnogonid living in another locality where the unsampled food may be more plentiful.

The feeding of the larval and post-larval immature forms of the pycnogonids may or may not be similar to that of the adults. In species with little yolk in the eggs the larvae feed independently earlier than in those species which have eggs with plentiful yolk. In the latter the larvae are retained on the ovigers of the male for considerable periods until development has progressed considerably further than in the species with eggs containing little yolk. Evidence to date suggests that several patterns of larval feeding exist.

(1) The larvae feed in a manner similar to that of the adults. This occurs in most members of the Ammotheidae.

(2) The method of feeding in larvae of members of the Colossendeidae and Tanystylidae is unknown.

(3) The larvae feed in or on hydroids although the adults have a different food. This occurs in members of the Endeidae, Phoxichiliidae and Pycnogonidae.

(4) Some larvae develop inside galls which form on the host animal but there is no evidence available concerning the mode of development or stimulus for formation of the galls. This has been recorded in the pycnogonid *Anoplodactylus exiguus* feeding on *Coryne muscoides*, *Anoplodactylus petiolatus* feeding on *Syncoryne eximia* and *Phoxichilidium femoratum* feeding on *S. eximia* or *Lucernaria* sp. These examples show that there is no great degree of specificity in which hosts can be induced to produce galls.

(5) The larvae develop to an advanced stage inside the eggs

relying upon the yolk to provide nutriment for growth. This occurs in some species of the Nymphonidae and Pallenidae.

More work is needed upon aspects of feeding in both the adults and the larvae. Existing information regarding observed feeding or association is summarised in Tables 3, 4 and 5 and from these the paucity of the information is obvious. Present records show that littoral species of pycnogonids feed mainly on hydroids, polyzoans or sponges in about that order of frequency but there is no evidence to suggest that any pycnogonid species is restricted to a single host species and they should be regarded as predators of a suitable range of host species depending upon a number of considerations. First, whether or not the pycnogonid can attack the prey without encountering the defensive or offensive mechanisms of the prey which include closing operculums or avicularia of polyzoans or the tentacles and nematocysts of hydroids. A second consideration is whether or not the pycnogonid can place itself in a stable position to attack the prey, which is important for the insertion of the proboscis through an integument, or for pulling away a hydroid head. A third is whether the proboscis is of the right shape and dimensions to attack the food.

Although little information is at present available it is possible that sponges or detritus are important food sources for species of pycnogonid living in deeper water, but in common with many aspects of pycnogonid biology much more information is needed regarding the feeding of these animals and their importance in the ecosystem.

There is not much information regarding predators of pycnogonids though they seem to form a small percentage of the food of the large Antarctic isopod, *Glyptonotus antarcticus,* an anemone of the Californian coast will eat *Pycnogonum stearnsi* and *Nymphon hirtum* has been found in the stomachs of fishes. It is extremely doubtful, and indeed highly unlikely, that pycnogonids provide a major food source for another animal except perhaps in deep waters.

Table 3
Records of larval associations

Pycnogonid species	Site of development	Source
Achelia alaskensis	In hydromedusae of Polyorchis karafutoensis	(Okuda, 1940)
Ammothea sp.	Galls in Coryne sp.	(Allman, 1862)
,, ,,	On nudibranch Armina varidosa	(Ohshima, 1933)
Lectythorhynchus hilgendorfi	On Holothuria lubrica	(Ohshima, 1927b)
L. marginatus	In Aglaophenia latirostris	(Ricketts and Calvin, 1960)
Endeis spinosus	On Obelia medusae and hydroids	(Lebour, 1916)
Nymphonella tapetis	In Paphia philippinarum	(Ohshima, 1927a)
Nymphon brevicandatum		(Fage, 1949)
N. hirtipes		(Hedgpeth, 1963)
N. robustum	Develop in egg	(Hedgpeth, 1963)
N. sluiteri	Carried by parent until sub-adult	(Hedgpeth, 1963)
Pallene sp.		(Fage, 1949)

N. parasiticum	On opisthobranch Tethys leporina	(Ohshima, 1927a)
Anoplodactylus erectus	Tubularia	(Hilton, 1934; Ricketts and Calvin, 1960)
A. exiguus	Podocoryne carnea or gall on Coryne muscoides	(Cole, 1904; Sanchez, 1959)
A. petiolatus	Galls on Syncoryne eximia	(Dogiel, 1911)
	In medusae of Cosmetira pilosella, Turris pileata, Stomotoca dinema, Phialidium hemisphericum and Obelia	
	Inside polyps of Hydractinia echinata, Podocoryne carnea and Obelia	
A. pygmaeus	In Obelia polyps	(Helfer and Schlottke, 1935)
Phoxichilidium femoratum	Forming galls in Syncoryne eximia, Lucernaria	(Helfer and Schlottke, 1935; Lebour, 1947)
P. tubulariae	Tubularia larynx	(Lebour, 1947)
P. virescens	Coryne muscoides	(Utinomi, 1954)
Pycnogonum littorale	Attached to Clava multicornis with proboscis inserted	(Dogiel, 1911)

Table 4

Associations which have been recorded between species of pycnogonids and other animals

Pycnogonid species	Association	Source
Achelia chelata	Inside Mytilus californianus	(Benson and Chivers, 1960)
Achelia echinata	Flustra foliacea	(Wyer and King, in prep.)
Achelia gracilis	Eudendrium ramosum, Obelia marginata	(Cole, 1904)
Achelia nudiuscula	Obelia	(Ricketts and Calvin, 1960)
Bohmia chelata	Gorgonia flammea, alcyonarians, hydroids and polyzoa	(Barnard, 1954)
Lecythorhynchus hilgendorfi	On Holothuria lubrica	(Ohshima, 1927b)
L. marginatus	Aglaophenia latirostris	(Ricketts and Calvin, 1960)
Colossendeis proboscidea	Porifera	(Stephensen, 1933)
Endeis spinosus	Obelia dichotoma	(Cole, 1910)
Ammothea discoidea	Hydroids	(Hedgpeth, 1941)
Nymphon hirtipes	Coral Eunephthya	(Stephensen, 1933)
N. maculatum	Soft corals	(Stock, 1963)

N. parisiticum	Opisthobranch *Tethys leporina*	(Benson and Chivers, 1960)
N. robustum	*Umbellula encrinus*	(Stephensen, 1933)
N. rubrum	Tubularians and *Halichondria panicea*	(Giltay, 1928)
Hannonia sp.	*Audouinia australis* (polychaete)	(Stock, 1959)
Pallene empusa	Tubularian hydroids	(Wilson, 1880)
Phoxichilidium femoratum	*Syncoryne* and *Aglaophenia latirostris*	(Lebour, 1947; Ricketts and Calvin, 1960)
P. hokkaidoense	*Eudendrium annulatum*	(Utinomi, 1954)
P. tubulariae	*Tubularia larynx*	(Lebour, 1947)
P. virescens	*Sertularia* on *Ascophyllum*	(Lebour, 1947)
Halosoma viridintestinale	*Obelia*	(Ricketts and Calvin, 1960)
Anoplodactylus angulatus	*Sertularia* on *Ascophyllum*	(Lebour, 1947)
A. erectus	*Corymorpha palma* and *Tubularia*	(Hilton, 1934)
A. lentus	*Eudendrium*	(Cole, 1901; Cole, 1906)
A. pygmaeus	*Bowerbankia* and hydroids	(Lebour, 1947)
A. petiolatus	On hydroids	(Giltay, 1928)
Pycnogonum benokianum	Anemone	(Hedgpeth, 1949)

Table 4—*cont.*
Associations which have been recorded between species of pycnogonids and other animals—*cont.*

Pycnogonid species	Association	Source
P. littorale	*Clava multicornis, Actinia, Tealia, Metridium, Alcyonarium digitatum*	(Dogiel, 1911)
P. rickettsi	*Anthopleura xanthogrammica*	(Ziegler, 1960)
P. stearnsi	*Aglaophenia* sp., *Clavelina huntsmani, Anthopleura xanthogrammica, Metridium senile* and *Bunodactis elegantissimum*	(Fry, 1965)
Austrodecus glaciale	*Cellarinella foveolata*	(Fry, 1965)
Rhynchothorax australis	*Eudendrium tottoni*	(Fry, 1965)
Tanystylum anthomasti	*Alcyonium pacificum*	(Utinomi, 1954)
T. californicum	*Abietinaria* and *Aglaophenia*	(Hedgpeth, 1951)
T. intermedium	*Aglaophenia latirostris*	(Ricketts and Calvin, 1960)

Table 5

Records of pycnogonid species actually observed feeding on hosts

Pycnogonid species	Food	Method of feeding	Source
Pycnogonum littorale	Milne-Edwardsia loweni, Tealia crassicornis, Lucernaria Cucumaria frondosa, Metridium dianthus, Metridium senile, Cynthia	Insertion of proboscis into tissues of the host	(Bouvier, 1923; Hedgpeth, 1949)
P. rickettsi	Anthopleura xanthogrammica		(Ziegler, 1960)
P. stearnsi	Anthopleura xanthogrammica, Bunodactis elegantissima		
Rhynchothorax australis	Eudendrium tottoni		(Fry, 1965)
Achelia echinata	Flustra foliacea, Dynamena pumila	Proboscis into operculum	(Wyer and King, 1971)
Austrodecus glaciale	Cellarinella foveolata	Proboscis inserted through the frontal pore	(Fry, 1965)
Anoplodactylus lentus	Eudendrium ramosum	Food forced into mouth by chelifores	(Cole, 1906)

Table 5—*cont.*

Records of pycnogonid species actually observed feeding on hosts—*cont.*

Pycnogonid species	Food	Method of feeding	Source
Phoxichilidium femoratum	*Tubularia larynx, Dynamena pumila, Tubularia, Eudendrium, Clava, Syncoryne, Coryne* and *Bougainvillea*		(Bouvier, 1923; Giltay, 1928; Lebour, 1947)
Ascorhynchus corderoi	Sponge		(Marcus, 1952)
Nymphon leptocheles	*Campanularia*	Proboscis inserted in hydranth which held by chelifores	(Helfer and Schlottke, 1935)
Nymphon gracile	*Actinia equina, Dynamena pumila, Bowerbankia, Nucella* eggs	Chelifores transfer food to proboscis	(King and Crapp, 1971)
Achelia longipes	Red algae on *Flustrellidra*	Browsing with the proboscis tip	(Wyer and King, 1971)
Endeis spinosus	Debris at base of *Antennularia*		(Wyer and King, in prep.)

6

GEOGRAPHICAL DISTRIBUTION

A thorough understanding of the geographical distribution of the pycnogonids is at the moment impossible because of insufficient sampling in certain areas and some degree of confusion concerning taxonomy so that many species have been described as distinct; when specimens from a number of localities have been examined some will without doubt be found to be synonyms. Examples of this are already known in the family Ammotheidae, particularly amongst species belonging to the genus *Achelia*. *Achelia spinosa* occurs on the cold Atlantic coast of North America, Sagami Bay, the Bering Sea and Point Barrow and is possibly synonymous with *Achelia echinata* which occurs on the Atlantic coast of Europe. Several species of *Achelia* were described by Losina-Losinsky from material collected at Lavrenti Bay in the Bering Sea. Some descriptions were based on males, some on females and others on juveniles. Hedgpeth, whilst examining material from Point Barrow, all the specimens of which were considered to be *A. spinosa,* remarked that the males resembled *A. lavrentii,* which had been described from male specimens by Losina-Losinsky; whilst the females resembled *A. uschakovi,* which had been described from female specimens; and the juveniles resembled *A. litke* and its variety *intermedia* which were based on juvenile material (Hedgpeth, 1963). This confusion reflects the taxonomic problems presented by the pycnogonids. Those species particularly which are restricted to a zone and have very poor locomotory ability must have a tendency to isolation and speciation.

In marine biogeography, the oceans of the world are usually divided into a number of zones, which may be termed Arctic, boreal, warm waters, antiboreal and Antarctic (Friedrich, 1969). In some schemes the terms differ but the areas are usually similar. Bipolarity, in the sense of identical species occurring in the Arctic and Antarctic regions, does not exist in the pycnogonids except in the case of ubiquitous cosmopolitan species such as some belonging to the genus *Colossendeis,* which are found in deep water in all oceans (Hedgpeth, 1947).

The pycnogonids of the Arctic region fall into three patterns of distribution (Fig. 28). First, there is a boreal-Arctic type, e.g.

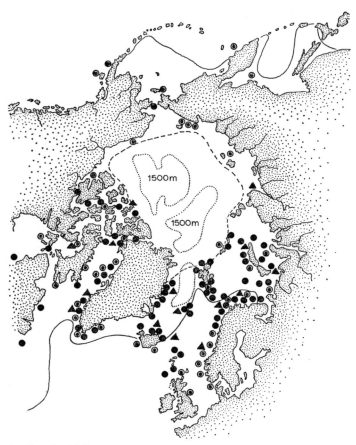

- *Nymphon hirtipes*
- ⊚ *Nymphon longitarse* (circumpolar distribution)
- ▲ *Colossendeis proboscidea*

Fig. 28

Examples of the patterns of distribution shown by Arctic pycnogonids. (Redrawn from Hedgpeth, 1963)

Geographical distribution

Nymphon longitarse, Nymphon grossipes and *Pseudopallene circularis* (Fig. 28). A second type of pattern can be recognised for species with abundant occurrences in deeper water of the boreal zones. Examples of species with this type of distribution are *Nymphon robustum, Colossendeis proboscidea, Nymphon hirtipes* and *Nymphon sluiteri* (Fig. 28). The third type of distribution is the north Pacific boreal pattern (Fig. 29) which is still incompletely known, but is characterised by such species as *Achelia borealis* and *Tanystylum anthomasti*. Only species in the first category are circumpolar in distribution. Evidence available at present suggests that the pycnogonid faunas of the two great boreal-to-Arctic regions of the Atlantic and Pacific are separate and have undergone independent evolution (Hedgpeth, 1963).

The fauna of the Sea of Okhotsk is peculiar and it has been suggested that it constitutes a separate province of development. In this area 40 per cent of the species are endemic (Fig. 29).

A number of deeper water species are found along the coast of west Greenland or in the Davis Strait–Baffin Bay area but no further westward. These include *Pallene acus, Pallenopsis calcanea* and *Nymphon leptocheles*. They probably arrived in this area via the Davis Strait. There are no species of the genus *Phoxichilidium* or *Pycnogonum* present. No species of *Pycnogonum* has been collected beyond Murmansk in the east, or as far as west Greenland in the west, across the Arctic Ocean. Distribution records suggest that the Bering Strait made very little contribution to the pycnogonids of the Arctic. The general scheme of provinces described for the pycnogonids (Fig. 29) does not agree with distribution patterns of other animals such as bivalves, though when more information is available, perhaps they will approximate more closely. Because of the difficulties involved in sampling, there is no information available concerning what species occur in the two deep Arctic basins on either side of the Lomonosov Ridge. It has been suggested that the pycnogonid fauna there will include the classic deep water species, *Nymphon robustum* and *Colossendeis proboscidea* (Hedgpeth, 1963). In contrast to the Arctic region which consists of an ocean, the Antarctic consists of a land mass surrounded by water. The northern limit of the region is usually considered to be either the Antarctic convergence or the subtropical convergence. The convergences are mainly surface features and the deeper layers of the ocean are continuous with the deep circulation of the major ocean basins to the north. The shape of Antarctica, which is almost circular, coupled with the continuous ring of water surrounding it, and the prevailing westerly winds, lead to the development, in most of the Antarctic Ocean, of a system of easterly currents, which form the great West Wind Drift

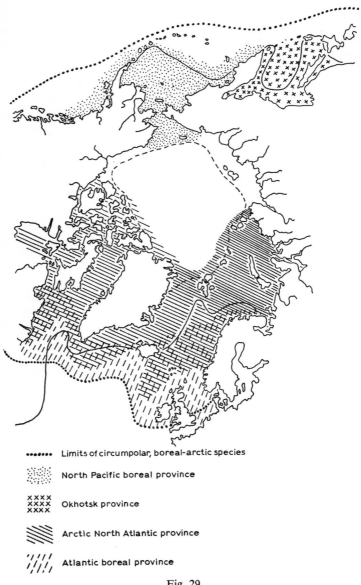

•••••• Limits of circumpolar, boreal-arctic species

North Pacific boreal province

xxxx
xxxx Okhotsk province
xxxx

Arctic North Atlantic province

Atlantic boreal province

Fig. 29

A general scheme of the pycnogonid provinces of the Arctic and boreal regions. (Redrawn from Hedgpeth, 1963)

Geographical distribution

(Fig. 30). Deep observations of temperature, salinity and distribution of elements such as oxygen, show that the whole body of water has this easterly movement, though the flow lines are influenced by the bottom topography and the structure of the current is extremely complicated (Fig. 31). Near to the land, there is a narrow region in which easterly winds prevail and where the water flow is westward, forming the East Wind Drift, which is contrary to the movement of the main body. When the coastal current meets the east coast of the Antarctic peninsular it is deflected northwards and flows eastwards across the Atlantic Ocean as the Weddell Drift. The boundary of the East and West Wind Drifts is the zone called the Antarctic Divergence. This is characterised by an upwelling of subsurface

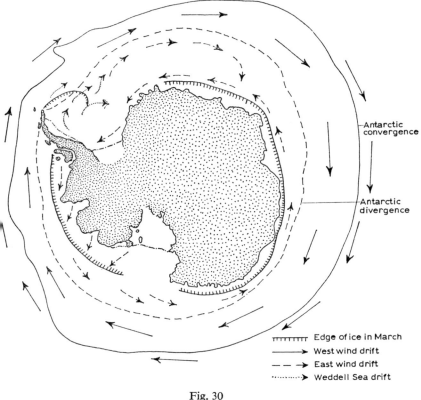

Fig. 30

Map of Antarctica showing the position of the Antarctic convergence and divergence and the surface currents.

water. The position and extent of the divergence depends upon the meteorological conditions prevailing at the time (Fig. 31).

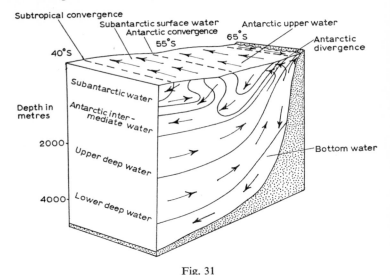

Fig. 31
Diagram of the meridional and zonal flow in the Southern Ocean.

Many of the genera of Antarctic pycnogonids are circumpolar (Fig. 32), and it is likely that when more information is available more will be shown to be so. This is true of *Colossendeis, Pycnogonum, Achelia* and *Ammothea*. In general, the degree of local endemicity on the Antarctic continental shelf is low, the circumpolar element forming the largest faunal component. There are twenty-eight species in the Ross Sea and only one subspecies is endemic (Fry and Hedgpeth, 1969).

The Magellanic region appears to be faunistically more discrete than other areas in Antarctica. Two monospecific genera, *Ascorhyncus* and *Anammothea*, are known only from this region which is the Antarctic Peninsula, and islands of the Scotia Arc and their surrounding shelves. The 2°C isotherm between South Georgia and the Falkland Islands, which is considered to be important in limiting the migration of asteroids and ophiuroids across Drake Passage, does not appear to be limiting for pycnogonids although it, or a similar temperature or salinity barrier, is undoubtedly important (Hedgpeth, 1947) (Fig. 33).

The Magellanic region provides the Antarctic pycnogonid fauna with a link to more northern shelf regions. Migration over the sea

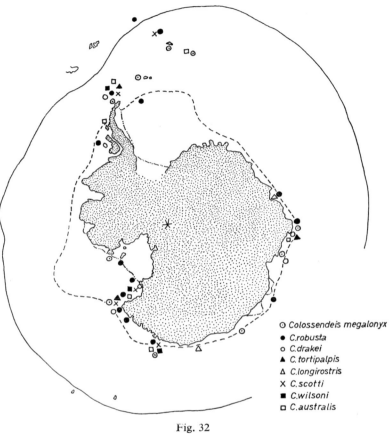

Fig. 32

Map showing the circumpolar distribution of the genus *Colossendeis* in Antarctica.

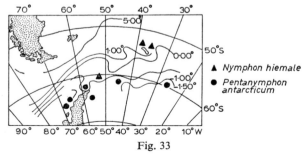

Fig. 33

Distribution of *Nymphon* and *Pentanymphon* compared with the temperature of 100 m surface.

bottom appears to occur extensively only in species belonging to the genus *Colossendeis*.

The genus *Achelia,* which has little locomotory ability, has undergone speciation here as in other parts of the world. Only five species of *Achelia* occur on the Antarctic continental shelf, and only one, *A. sufflata,* is confined to the Antarctic continent. This species has close morphological similarities with some South American species. A number of morphogroups exist amongst the species occurring in the Antarctic. In this context a morphogroup is a number of species which form a morphologically homogeneous group, isolated by the morphology of its components from other groups or single species. One morphogroup consists of *A. communis, A. spicata, A. serratipalpis* and *A. hoekii.* The first two of these have a circumpolar distribution. A number of groups of closely related species sometimes referred to as morphogroups are apparent amongst the circumpolar species, *A. communis, A. spicata, A. serratipalpis* and *A. hoekii* (Fry and Hedgpeth, 1969) (Fig. 34).

A second group of species forms a South American morphogroup consisting of *A. parvula, A. fernandeziana, A. sufflata* and *A. besnardi.* The latter has also been recorded from the Gulf of Guinea and the eastern coast of South America. No information is available concerning the frequency of this occurrence and northern surface currents in the south Atlantic may account for these records. A central American morphogroup is represented by *A. sawayai* and *A. gracilis,* both of which occur north of the equator (Fig. 34).

The south temperate morphogroup consists of *A. australiensis* which occurs around New Zealand and the southern coasts of Australia, *A. dohrni* which is restricted to New Zealand and *A. assimilis* which occurs in New Zealand and South America (Fig. 34).

Excluding *Colossendeis,* the Magellanic region has five genera in common with South America. In this instance there is only one species, *Ecleipsothremma spinosa,* which is common to both areas.

Although there are no fossils, there is some evidence of fairly recent migrations from the Magellanic region to South America and Australia. This evidence is the morphological similarity between *Ammothea magniceps* of Australia and *A. minor* and *A. clausi* of Antarctica. This suggests a migration by the Westward Drift (Fig. 30). The similarity of *Achelia sufflata* to other species in South America indicates movement across Drake Passage.

The general distribution of the pycnogonids agrees in outline with that of other invertebrate groups that have undergone active speciation in the Antarctic. The region would be ideal for this to occur because of the dominance of sessile animals such as bryozoans,

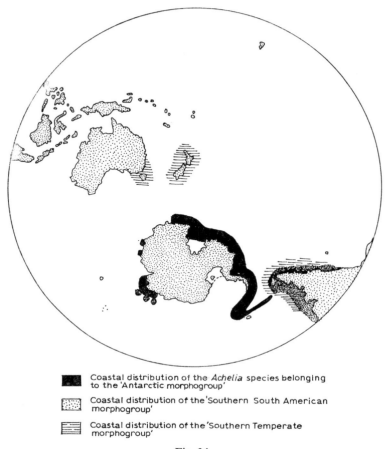

Fig. 34
Distribution of the morphogroups of the species of the genus *Achelia* around the Southern Ocean. (Modified from Fry and Hedgpeth, 1969)

hydroids and sponges in the benthic zone, which are believed to provide the chief food source for pycnogonids (Fry and Hedgpeth, 1969).

Although there are many small genera in scattered parts of the tropics, there is only one large genus, *Anoplodactylus,* which may be considered as typically tropical. This genus is represented by several species in temperate latitudes but has its greatest number of species in the tropics, particularly in the West Indies.

The pycnogonid fauna of Japan, although incompletely investigated, has a large endemic element. Part of this element is represented by the genus *Ascorhynchus* which has six well-defined species in Japanese waters and a number of species belonging to the genus *Nymphon* (Stock, 1954). No endemic genera have been found. It was once thought that *Nymphonella* was endemic but it has subsequently been found in the Mediterranean. The pycnogonid fauna of this region is not uniform but consists of a southern, subtropical fauna and a northern, cold-temperate fauna. The boundary between these two faunas is situated at about latitude 35°N. The southern Japanese pycnogonid fauna agrees in many respects with that of the East Indies. This is especially evident in the distribution of *Nymphopsis muscosa* and representatives of the genus *Pallenopsis* which is mainly a warm-water genus. No representatives of this genus occur north of latitude 36°N in Japanese waters, but south of this there are six species of the genus and five of these, *P. tydemanni, P. sibogae, P. molissima, P. virgatus* and *P. temperans,* also occur in the East Indian region (Stock, 1954). The northern fauna contains several boreal-Arctic species. The fauna of the Japanese region is markedly different from that of the eastern Pacific, one notable absentee being members of the genus *Tanystylum*. This divergence between east and west perhaps results from the intrusion of masses of Arctic water through the Aleutian chain and the absence of a convenient transport system in the form of Sargassum weed.

There are a few species which occur in Japanese waters and also off California. These are *Decachela discata, Pycnogonum stearnsi, Ammothella biunguiculata* and *Lecythrorhynchus marginatus*. Another species, *Achelia latifrons,* occurs off California and Korea.

The East Indian fauna is strongly characterised by a number of endemic genera; *Scipiolus, Hemichela, Pycnofragilia* and *Pycnopallene*. The Indo-Malayan region has a number of species in common with India, *Propallene longiceps, Pseudopallene hospitalis, Pallenopsis ovalis, Anopladactylus saxatiles, A. grandulifer, Endeis meridionalis, E. flaccida* and *Rhopalorhynchus kröyeri* (Stock, 1954). The relationship with the southern Japanese pycnogonid fauna has already been mentioned. At a generic level there is also some relationship with the fauna of the West Indies, Caribbean and Bermuda. This is shown by the distribution of *Heterofragilia, Parapallene* and *Ascorhynchus*. The last two genera are mainly confined to the Indo-West Pacific region except one species of *Parapallene, P. bermudensis,* which has been collected from Bermuda and one species of *Acorhynchus, A. armatum,* also from the West Indies. There is a slight relationship with the fauna of South Africa and a small one with tropical and subtropical Australia (Stock, 1954). Only three species are at present known to be common to both faunas: *Ascorhynchus melwardi* has been recorded from Singapore and the Torres Straits, *Pallenopsis hoeki* is fairly common in both regions and *Rhopalorhynchus kröyeri* is widely distributed in the Indo-West Pacific and as far south as the Queensland coast in Australia. The Australian region appears to be characterised by an abundance of genera and species particularly in the Pallenidae, many of which are endemic or appear to have a centre of distribution in the Australian region or in the Austro-Malayan region. An interesting type of distribution is that shown by the species of *Oropallene* and *Anoropallene* with *Oropallene* being mainly Australian in distribution and *Anoropallene* mainly restricted to the Pacific coast of North America. *Anoropallene valida* from Australian waters appears to be transitional between the two genera (Clark, 1963).

The New Zealand fauna shows a high degree of independence with ten endemic species present in the total of seventeen littoral species occurring on those shores. There are, however, at present no endemic pycnogonid genera known from the coastal waters of New Zealand. The genus *Oorhynchus,* the only species of which, *O. aucklandiae,* has been taken only once off Auckland, is a deep water species, and probably will have a much larger distribution. Some relationship exists with the fauna of south Australia and Antarctica (Stock, 1954). The pycnogonid fauna of the temperate regions of the north Pacific which comprises the faunas of the Pacific coast of America from lower California to the Gulf of Alaska and the coast of Japan, north of latitude 36°N, seems to be much richer than that of the north Atlantic. The north Atlantic has only one endemic genus, *Paranymphon,* a deep water form, whereas there are at least three endemic genera in the north Pacific, *Lecythorhynchus, Nymphonella* and *Decachela*. One notable difference between the pycnogonid faunas of the two oceans is the relatively small number of species which are widely distributed along both shores of the Pacific. Exceptions to this are *Lecythorhynchus marginatus* and *Decachela discata*. This contrasts with the relatively large number of littoral species which occur on both sides of the Atlantic. Few of the boreal-Arctic species occur in the north Pacific. They are represented by *Nymphon grossipes, N. longitarse* and *Phoxichilidium femoratum*.

The faunas of the Pacific and Atlantic coasts of tropical and subtropical America have several common traits in spite of the fact that the Isthmus of Panama forms a faunistic barrier for marine animals. There is a closer relationship between the faunas than would be expected from the present isolation from each other. At least four species occur in both areas: *Nymphopsis duodorsospinosa,*

Tanystylum calicirostre, Pycnogonum reticulatum and *Anoplodactylus portus* (Hedgpeth, 1948; Stock, 1955). In addition to these a number of closely related or twin species, one from the Pacific coast and one from the Atlantic coast, have been described which may become differentiated after the isolation of the two areas by the rising of the Isthmus. Listing the Atlantic forms first, these pairs are *Callipallene emaciata/C. californiensis, Ascorhynchus armatus/A. agassizi, Eurycide raphiaster/E. longisetosa, Ammothella regulosa/A. heterosetosa, Tanystylum orbiculare/T. duospinum* (Hedgpeth, 1947) and *Tanystylum gemium/T. isthmiacum* (Stock, 1955). The Caribbean has an extremely rich pycnogonid fauna which includes two decapodous species and the genera *Ephyrogymna, Pentacolossendeis, Neonymphon* and *Cheilopallene*, which are endemic. The pycnogonids are poorly represented on the Pacific side of the Isthmus with no endemic genera and only a few species of such widely distributed genera as *Tanystylum, Anoplodactylus* and *Pycnogonum*. The West Indian fauna shows a distinct affinity to that of the eastern Atlantic, which includes the coasts of Europe, Mediterranean and the western coast of north Africa (Fig. 35). There are twenty-three species known at present which occur in both regions. Considering the evidence, Stock (1955) stated that the West Indies must have had, at an early date, a connection with the Indo-West Pacific, as suggested by the considerable number of joint genera. Regarding species, however, the West Indies have a greater resemblance to West Africa than to the Indo-West Pacific. This implies that the separation of the West Indian and Pacific faunas by the rising of the Isthmus is a fairly recent one.

To summarise the knowledge on the geographical distribution of pycnogonids, there are at least twice as many endemic genera in the southern as in the northern hemisphere and most of these are widely distributed and circumpolar. In the northern hemisphere, all except one deep-water form of the endemic genera are restricted to the Pacific, excluding the tropical forms. Normally, distribution of pycnogonids has depended upon prevailing currents, contiguous ocean shelves, or floating material such as Sargassum (Fig. 35). If the distribution is tentatively considered in these terms, a pattern begins to emerge, but it must be stressed that much more data are needed before any pattern can be conclusively elucidated.

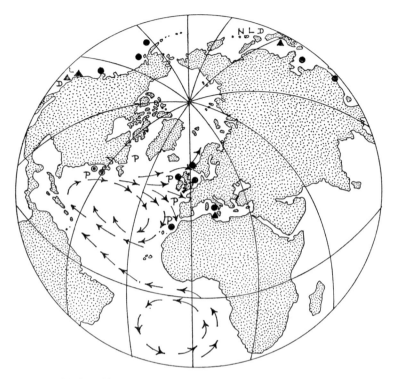

● *Achelia echinata*
⊚ *Achelia spinosa*
← Atlantic currents
▲ *Ammothella bi-unguiculata*
 (Records at Hawaii and Western Australia not shown)

Distribution of endemic species:
D - *Decachela*   L - *Lecythorhynchus*
N - *Nymphonella* P - *Paranymphon*

Fig. 35

Map showing currents and possible directions of dispersal in the Atlantic Ocean originating in the West Indies and the distribution of several endemic species. (Redrawn from Hedgpeth, 1947)

7

AFFINITIES AND EVOLUTION

It has been suggested that several fossils represent ancestral pycnogonids but considerable doubt exists concerning this relationship (see page 129) so that any statements regarding the origins and affinities of the pycnogonids must remain as pure speculation. When comparisons are made between pycnogonids and crustaceans, arachnids or other arthropods the evidence concerns living forms only and the degree of similarity with ancestral types is, in many instances, uncertain.

Opinions regarding the relationships of the pycnogonids usually support one of three possibilities:

(1) That they are more closely related to the crustaceans than to other arthropods. Evidence cited in favour of this is some external resemblance between the nauplius larva of crustaceans and the protonymphon of the pycnogonids; a similarity between the process of vitellogenesis during oogenesis in the two groups; a similarity between the process of gastrulation in some pycnogonids and isopods; both moult after the adult stage is reached; both have representatives with an anamorphic type of development; a hermaphrodite has been recorded in the pycnogonids and some occur in the *Cirripedia* and *Cephalocarida* amongst the Crustacea; both are predominantly marine and representatives of both brood or carry the young. However, many of these characteristics could easily have been developed independently of the two groups so that they do not provide strong evidence concerning a relationship between them.

(2) That they are more closely related to the arachnids than to other arthropods. Evidence cited in favour of this is that both have

caeca of the gut entering the limb bases, a resemblance between the process of embryogenesis in *Callipallene* sp. and some arachnids, the eyes of pycnogonids and most chelicerates are of the simple type and are situated on tubercles in pycnogonids and some Araneae and Phalangidae, the only pair of pre-oral appendages in both groups are the chelifores and chelicerae respectively, and both groups lack a deutocerebrum in the cephalic ganglionic mass. Many of these characters also have doubtful validity as evidence for a relationship between the pycnogonids and the arachnids.

(3) The view now more generally held is that the pycnogonids should have the status of a class or subphylum. This implies an earlier breakaway from the other arthropods and consequently a different set of relationships with them. Evidence in favour of this, which is mainly concerned with refuting the ideas of the other two alternatives, is extensive. The ovigerous legs of the pycnogonids have no counterparts in the arachnids. In those chelicerates which have been studied in detail, the germ primordia appear early in development, and the position of the genital openings is remarkably constant, being situated on the second abdominal segment. It is hard to envisage a forward migration of the gonad from the abdomen which would result in the development of genital openings on all body somites. Some workers consider that the segmentally arranged genital openings are a primitive feature only found, apart from the pycnogonids, in the polychaetes. Similarly, they consider that the gut diverticula resemble those in polychaetes (Sharov, 1966).

The anal appendage or abdomen of the pycnogonids may represent all that is left of a larger abdomen in the ancestors. This phenomenon of reduction of segmentation progressively from the posterior end is thought to be widespread in arthropods. Since it has reached an advanced stage in pycnogonids, it would suggest that they are an ancient group (Hedgpeth, 1964). In the absence of a fossil record, however, all views on the subject will remain as speculation on the most tenuous of evidence.

The eye tubercle is claimed to resemble the unpaired tentacle of the aboral complex in the polychaetes. If originally present, the gill outgrowths on the limb were probably lost very early in the evolutionary development of pycnogonids. During cleavage, some pycnogonid eggs have a very rudimentary type of spiral cleavage which also occurs in the annelids though it is believed to have been lost by the arthropods very early in their evolution (Sanchez, 1959).

It is possible that the pycnogonids originated very near to the basic Arthropodan stock and are perhaps related to a form resembling the annelid *Spinther* rather than the other living annelids. The fact that pycnogonids feed mainly on hydroids and sponges has been cited as evidence for their ancient origins (Sharov, 1966).

The cuticle of pycnogonids, in common with that of annelids and tardigrades, has an outer film of mucus. This may be a secondary adaptation, but if not, it supports the idea of an early breakaway from the arthropod stock.

The proboscis, if it is the homologue of the annelid proboscis, presumably lost the ability to retract and associated with this the labrum would presumably be lost. Observations, based on the innervation of the proboscis, assuming that the pycnogonids have followed the same evolutionary pattern as other arthropods, suggested that the stomodeal system of the pycnogonids is directly homologous with that system in other members of the Annulata. Furthermore, it was suggested that the proboscis is composed of elements of the first and second body segments and is thus homologous with the labrum of the Crustacea, the clypeus of the insects, the rostrum of *Limulus* and the proboscis of the Polychaeta. The segments or antimeres of the proboscis do not represent fused segments. A number of other possible interpretations of the segmentation of the body in pycnogonids have been made and since all opinions are merely speculation virtually none can be ruled out. Regarding the theory expressed above, Hedgpeth (1964) has pointed out that, considering some abnormal specimens of pycnogonids which have been recorded, care must be exercised in stating dogmatically that a nerve from any particular ganglion will always serve the organs of the same segment. Also, since the annelid proboscis consists of two annular units, why should the proboscis of the pycnogonids have three longitudinal antimeres? During embryological development, the primary appendages appear later than the proboscis and this structure is usually situated well back on the ventral surface. The question can be posed in return, however, why should the proboscis of living polychaetes resemble that of the parent stock of the arthropods any more than that of the pycnogonids? Both groups as stated earlier must be considered as survivors of a long evolutionary process and the pycnogonid proboscis together with many other characters may have evolved after their divergence as an independent group from the basic stock.

In the absence of a fossil record, the problem of pycnogonid origins and evolution cannot be satisfactorily solved but, to sum up the existing evidence, it is probably true to say that they are a class of animals not closely related to any other group of arthropods, which suggest that they arose early and have retained a mixture of primitive together with specialised characters which has led them to be described as representing an anthology of the arthropods.

8

SYSTEMATICS

There are without doubt many more species of pycnogonid to be discovered, particularly from deeper waters, but to date approximately 500 have been described, are distributed amongst 8 families and 70 genera; and more than 100 of them belong to the genus *Nymphon*. Attempts to erect groups higher than the family level fail because of the large number of intermediate forms which render these groupings invalid. The most satisfactory arrangement is that all the living species should be assigned to one order, the Pantopoda (Gerstaeker), and all the fossil species to another order, the Palaeopantopoda (Broili) (Hedgpeth, 1947). A thorough review of pycnogonid taxonomy is needed but a compromise classification, though not completely satisfactory, divides the order Pantopoda into eight families: the Colossendeidae, Pycnogonidae, Endeidae, Nymphonidae, Ammotheidae, Pallenidae, Phoxichilidiidae and Tanystylidae (Hedgpeth, 1947).

The families Endeidae, Pycnogonidae and Colossendeidae each consists of groups of closely related genera and thus are reasonably well-defined families. The families Ammotheidae and Pallenidae are of very questionable status, since each consists of a group of genera with a wide range of structural variation. A number of forms which are intermediate between families exist (Figs. 36, 38c,f). *Rhynchothorax* is considered to be intermediate between the families

Pycnogonidae and the Tanystylidae (Figs. 38a,e, 40a). *Pigrogromitus* (Fig. 38c), a form recorded from the Suez Canal, has the general body shape which is characteristic of the Pycnogonidae but has ten segmented ovigers which are typical of the Pallenidae (Fig. 39d). *Pseudopallene,* which is not as elongate as the other members of the Pallenidae, is perhaps an intermediate between this family and the Tanystylidae (Hedgpeth, 1947). *Pallenopsis* has ten segmented ovigers but these are reduced in the females a character which links them with the Pallenidae. In addition to this possible relationship they have cement glands on the femurs of the males and an overhanging cephalic segment, both characteristics suggesting a link with the Phoxichilidiidae (Hedgpeth, 1947).

Hedgpeth (1947) suggested that the pycnogonid stock is best represented diagrammatically as a hemisphere with the relationships between the families indicated by lives in two dimensions. A possible relationship is shown in Fig. 36.

ORDER PANTOPODA

Family Colossendeidae (Figs. 37a–m)

The characters of this family are: a trunk with usually little outward visible segmentation and palps consisting of eight or nine segments which in the adults are always as long as or longer than the proboscis. This is large compared with the trunk, varying in different species between three-quarters the length of the trunk and twice the length of the trunk. Examples of this are shown by the relative proportions of the proboscis to the trunk in *Colossendeis tortipalpis* which varies between 1·62 and 1·7 and in *Colossendeis longirostris* it is 2·75. It is possible that *C. longirostris* is an essentially abyssal subspecies of *C. tortipalpis* and both of them are Antarctic species (Fry and Hedgpeth, 1969). In most species the widest parts of the proboscis are near the tip and near the middle whilst the diameter of the rest of the proboscis is similar to that of the trunk. Notable variations from this pattern is the proboscis of *Pipetta weberi* in which the distal two-thirds is very narrow and of *Rhopalorhynchus kröyeri* in which the proximal half of the proboscis is very narrow. Little is known concerning the food of members of this family except that in the Antarctic, *Colossendeis lilliei* and *Colossendeis megalonyx* have frequently been collected associated either with polyzoans or with a sponge-hydroid complex and *Decalopoda australis* associated with *Tubularia* (Fry and Hedgpeth, 1969).

The ovigers which are present in both sexes have ten segments. The four terminal segments of the palp have patches of spines on their ventral surfaces. These spines are frequently smaller on the

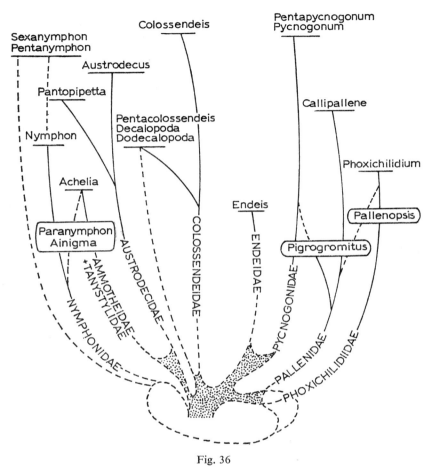

Fig. 36
A speculative diagram showing some of the possible relationships between families. The possible position of *Paranymphon*, *Ainigma*, *Pigrogromitus* and *Pallenopsis*, which are sometimes considered as transitional forms, is indicated. The Tanystylidae and Ammotheidae are considered a single family, while the Austrodecidae are considered by some workers to be of family status.

ovigers of the female than on those of the male. The shapes of these spines fall into four main categories which are called peg, needle, spatulate and molariform (Fry and Hedgpeth, 1969). Some species have only one or two types of spines on their ovigers but all types of spine are present on the ovigers of some species belonging to the genera *Colossendeis, Decalopoda* and *Dodecalopoda*. The spines may be arranged either as an undifferentiated field of peg and needle types or an undifferentiated field together with a single row of larger spines as in *C. tortipalpis*. The functional significance of this variation is unknown and the absence of palaeontological data makes it impossible to speculate on the possible phylogenetic significance of these variations. The terminal segment of the oviger has a larger spine than the other segments. Observations on Antarctic species have shown that the ovigers are used in both sexes for grooming to keep the surfaces of the body and appendages clear (Fry and Hedgpeth, 1969). It has been suggested that both the *Decalopoda* and *Colossendeis* may use the ovigerous legs in feeding, using them to rake the sea-bed.

There are between four and six pairs of legs depending upon the genus with *Colossendeis* and *Rhopalorhynchus* having four pairs, *Decalopoda* and *Pentacolossendeis* five pairs and *Dodecalopoda* six pairs. The legs are extremely long and thin, varying between 5 and 16 times longer than the trunk and they are without accessory claws. The tarsus is straight and cylindrical and has a length subequal to that of the propodus. The trunk segments of *Pantopipetta* have a diameter only slightly greater than that of the femora of the legs. In members of the *Colossendeidae* the trunk never has prominent dorsal ridges or projections although there is a prominent post-

Fig. 37

A range of the body forms in the family Colossendeidae. (Redrawn from Fry and Hedgpeth, 1969)
(a) *Colossendeis longirostris*.
(b) *C. stramenti*.
(c) *C. tortipalpis*.
(d) *C. stramenti*.
(e) *C. australis*.
(f) *C. megalonyx*.
(g) *Rhopalorhynchus kröyeri*.
(h) *Pipetta weberi*.
(i) *Decolopoda*.
(j) Terminal segments of the oviger of *C. megalonyx*.
(k) Ovigerous spines.
(l) Dodecolopoda.
(m) Chelifore of juvenile and whole specimen showing one palp and chelifore.

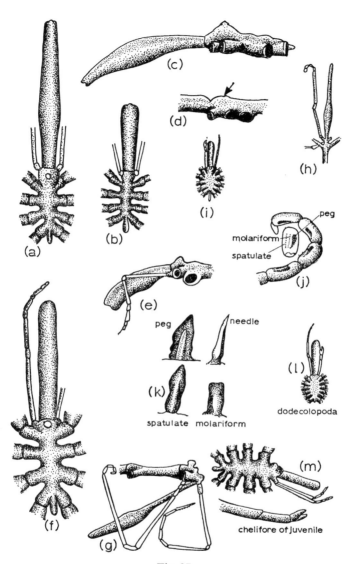

Fig. 37

ocular lump as in *C. tortipalpis* and there are no large projections on the lateral processes of the trunk or coxal segments of the legs. An ocular tubercle is usually present, though in *C. megalonyx orcadense* the eyes are sometimes absent or, if present, they are unpigmented (Fry and Hedgpeth, 1969). In *Colossendeis stramenti* the ocular tubercle is very low and rounded, being wider than long, and almost as wide as the cephalon with some suggestion of separation into separate lobes, but eyes are completely absent.

In the juveniles the chelifores are strongly chelate each with a two-segmented scape but in most species they are deciduous when the adult stage is reached. Chelifores are retained in the adults of *Decalopoda* and *Dodecalopoda* but are lost in the ten-legged *Pentacolossendeis* which occurs in the Caribbean.

In both sexes the genital pores are present on the ventral surface of the second coxae of all legs, those in the male being smaller and situated on a low protuberance whilst those of the female are larger but without a protuberance.

Reproduction and life histories of members of this family are unknown and the juveniles of only a few species have been described. No males have been found carrying eggs or developing larvae so doubt exists as to whether this occurs.

About thirty species of the genus *Colossendeis* have been described. Sixteen of these species have been found in waters south of 40°S and of these species, twelve are endemic to this region where most are usually confined to waters less than 1000 metres deep. The other four species are cosmopolitan and, in contrast with others, have usually been found deeper than 1000 metres and in some instances as deep as 5000 metres.

The group of endemic species appear to form a morphologically discrete group which are more closely related to each other than to the cosmopolitan species. The possession by the cosmopolitan, deep water species of an eye tubercle with atrophied eyes suggests that they may have been derived from shallow-water forms. Unless one postulates a vanished shallow-water fauna in other parts of the world the only coastal and shelf regions which support or have supported an endemic *Colossendeis* fauna is the Antarctic (Fry and Hedgpeth, 1969).

At the present state of knowledge regarding the taxonomy of this genus, the Antarctic *Colossendeis* fauna appears to be discrete, though shortage of knowledge regarding the cosmopolitan species prevents an accurate appraisal of the relationships between the two groups. Some of the cosmopolitan species appear to be stenothermous and eurybathic and have been taken in very shallow Arctic waters though they are confined below 1000 metres in the Antarctic

Systematics

and Subantarctic waters. This suggests that there are conditions peculiar to shallow and shelf regions of the Southern oceans which prevent the exploitation of shallower waters in the Antarctic by any but the specialised forms. However, the endemic shallow-water species seem to have been able to contribute to the bathyal and abyssal forms (Fry and Hedgpeth, 1969).

Family Pycnogonidae (Fig. 38a–b)

The characters of this family which is of world-wide distribution are: the absence of chelifores and palps in the adults; and ovigers which are 4- to 9-segmented with a large terminal claw present in the males only. Barnard (1954) described a specimen of *Pycnogonum portus*, presumed to be a male, which lacked ovigers, but carried a flat mass of eggs cemented to the ventral surface of the body. A similar occurrence was described in *Pycnogonum anovigerum*. The ovigerous legs first appear as small protuberances in a young larva and increase in size and number of segments at each moult (Fig. 8a). The propodus of the leg is well developed, but without a heel or heavy basal spines; the tarsus is arcuate and much shorter than the propodus. This family is separated from the Endeidae by the absence of accessory claws at the ends of the legs which are present in that family. Members of the Endeidae are generally fairly elongate and lightly built whereas the members of the Pycnogonidae are relatively thick-set in the body and the diameters of their legs are large in relation to the length of the leg. The trunk has a large diameter in relation to its length and the proboscis is usually stout and never longer than the trunk.

The genus *Pycnogonum* contains twenty-three species, at least one of which is present in every major ocean basin and in the Antarctic there are a number of species. Examples of these are *Pycnogonum gaini* which has an angular appearance and relatively long legs, *Pycnogonum rhinoceros* which has numerous prominent tubercles on the proboscis and a well defined patch of sharp spines at the base of the tarsus and *Pycnogonum platylophum* which is a relatively small, clean-limbed species with rather narrow trunk segments. It is possible that this last-named species is the same as *Pycnogonum magellanicum* though, as with so many other species, this point has not been elucidated (Fry and Hedgpeth, 1969).

The internal arrangement of organs in this family is similar to that in other families but there are some differences. In the female the reproductive system has the typical shape but egg maturation is not confined to a few segments of the legs but occurs also in the trunk. The genital openings are present only on the second coxae of the last pair of legs, which are the fourth pair in *Pycnogonum* and the

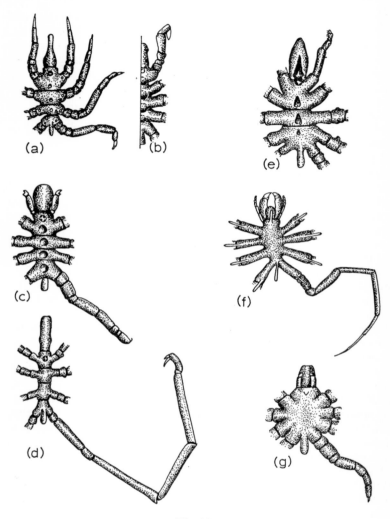

Fig. 38
The body form of the Pycnogonidae, Endeidae and some species of doubtful affinities.
(a) *Pycnogonum littorale.*
(b) *Pentapycnon.*
(c) *Pigrogromitus timosanus.* (After Hedgpeth, 1947)
(d) *Endeis spinosus.*
(e) *Rhynchothorax mediterraneus.* (After Dohrn, 1881)
(f) *Paranymphon spinosum.* (After Hedgpeth, 1947)
(g) *Decachela discata.* (After Hedgpeth, 1947)

Systematics

fifth pair in *Pentapycnon*. After release from the female, the eggs are carried by the male, usually as a single large mass in which the ovigerous legs are embedded. The body surface in most species has a warty appearance and in *Pycnogonum littorale* the warts are arranged in a series of rosettes over the surface of the body except for the cuticle covering the eyes and the apical third of the proboscis each with a central pore leading to the duct of mucous cells beneath the cuticle. The function of these cells has not been determined.

Family Endeidae (Fig. 38d)

The characters of this family are: the absence of chelifores and palps in the adults although the former are present in the juvenile stages, four pairs of walking legs, ovigers present in the males only with 7–8 segments. The tarsus is short and arcuate and the propodus is well developed and arched with a pronounced proximal heel and heterogenous sole spines. The legs are long and slender and at their distal ends have well developed accessory claws. The Endeidae is numerically a relatively small monogenetic family of world-wide occurrence. The members of this family are mostly shallow-water animals and two species have been recorded from floating seaweed and plankton tows. *Endeis spinosa* is probably a permanent member of the Sargassum fauna where it is thought to live amongst the colonies of the hydroid *Obelia dichotoma*. It is a common well-known species found on European coasts from Norway to the Black Sea and the Azores and also on the eastern coasts of North and South America and has on occasions been observed swimming. Little is known concerning the feeding habits in this family but the smaller, presumably immature, individuals usually have light green gut contents and the largest specimens, which are presumably mature, have red-coloured guts. This difference suggests that the diet of the adults and juveniles differs. Specimens have been observed feeding on the polyps of the hydroid *Tubularia larynx* in North Wales and it is believed that others feed on *O. dichotoma* (Crapp, 1968) or on detritus which gathers on the surface of *Antennularia*. *Phoxichilidium tubulariae*, which also feeds on *T. larynx*, grasps the hydranths with chelae when feeding but *E. spinosa*, which lacks chelifores, uses the first pair of walking legs to grasp the hydroid and hold it against the proboscis (Crapp, 1968). Lebour (1916) recorded the larvae from both inside the polyps of *Obelia* and also clinging to medusae in the plankton.

For more than half a century controversy has raged concerning the standing of the generic name *Endeis* and earlier references often called this genus *Phoxichilus*. In spite of its wide geographical range the genus is morphologically remarkably compact and homogeneous.

The separation of the species is based upon characters which in this genus show a very small degree of variation but tend to play only a minor part in the keys of other families (Fry and Hedgpeth, 1969).

Family Pallenidae (Fig. 39d)

The family Pallenidae consists of about 110 species forming a rather heterogeneous group. Many of these species have an elongate, slender build and are chiefly shallow-water animals, though a few species are found at depths greater than 1000 metres. There is evidence that some species live amongst hydroids, but some others are apparently planktonic, possibly associating with medusae. *Callipallene brevirostris* is a common shallow-water species, found from Scandinavia to the Mediterranean, and also on the Atlantic coast of North America. *Callipallene phantoma* occurs from Norway to Naples and the Azores, the Black Sea and the Florida coast. Opinions vary concerning the validity of many of the species and some workers believe that many at present believed distinct are synonyms (King and Crapp, 1971).

Family Nymphonidae (Fig. 39a–c,e,f)

The members of this large family are extremely uniform and many of its species are predominantly deep-water forms though some are often found in shallow water. The Nymphonidae usually have a slender body with a relatively short, wide proboscis. The chelifores are usually chelated, each of them being two-jointed and with their fingers frequently bearing prominent teeth. The palpi are five-jointed and the ovigerous legs, which are present in both sexes, are ten-jointed. If the Pycnogonidae are considered to be the most specialised family, because of the absence of palps and chelifores, then the Nymphonidae are probably the most generalised group, retaining more of the primitive attributes by having fully developed palps and chelifores. It is more likely, however, that these characters represent adaptations to the mode of feeding. Although of wide occurrence the Nymphonidae are particularly abundant in the North Atlantic. Several species have been observed swimming and they appear to be among the most active members of the pycnogonids, and are thus less likely than members of some other families to form local varieties. Correlated with the fact that they usually produce eggs which are rich in yolk, many species undergo development until they resemble the adult form with a full complement of appendages though much smaller whilst attached to the ovigers of the male parent.

There are three species which commonly occur in the littoral zone in the British Isles, *Nymphon gracile, N. rubrum,* and *N. brevirostre*

(King and Crapp, 1971). *N. rubrum* can be separated from the other two by having a higher length/width ratio of the third trunk segment (Fig. 39e) and *N. gracile* can be distinguished from the others on a comparison of the length of the ultimate and penultimate palp segments (Fig. 39f). In the polar seas of the USSR these species do not appear to be distinct and it is thought that *N. brevirostre* and

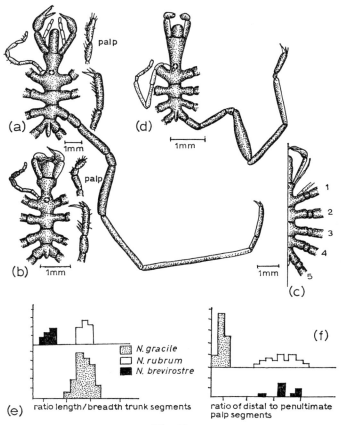

Fig. 39
The form of the body in the Nymphonidae and Pallenidae.
(a) *Nymphon gracile.*
(b) *N. rubrum.*
(c) *Pentanymphon.*
(d) *Callipallene brevirostre.* Histograms to show the characters on which the British species are distinguished.
(e) Ratio of length to the breadth of trunk segments.
(f) Ratio of distal to penultimate palp segments.

N. rubrum may be synonyms. All the specific characters transgress and there is a pronounced variability of the age character and the different stages of development in the separate forms which may be similar to each other. It is possible that the open-sea forms, which live at great depth on slime bottoms, grow to a larger size than the coastal ones which settle in the littoral or immediate sublittoral zones on different bottoms with a rich fixed fauna. The open-sea forms are said to retain the characters peculiar to the young individuals of the species. It is thought that the variation could be the result of either a stunted development of the post-larval stadium, differential acceleration of some characters or an unequal ratio of growth in some parts of the animal. This is only a small example of the type of problem facing taxonomists trying to sort out the pycnogonids.

Most species have four pairs of walking legs, but one genus has five and another six.

Family Phoxichilidiidae (Fig. 40c,d)

This is, considering species number, a fairly small family consisting of about fifty-three species. Many of them occur closely associated with hydrozoans and seven species are known to have larvae which develop parasitically in or on hydroids or hydromedusae. Both adult and larval stages are occasionally taken in plankton hauls. The characteristics of this family are that the chelifores are chelate, the palps rudimentary, or, as is more usual, absent and the ovigerous legs are present only in the male (Fig. 40). *Phoxichilidium femoratum* is common in the north Atlantic and Pacific Oceans and *Anoplodactylus petiolatus* occurs on both sides of the Atlantic. Other species occur in shallow water in the tropics and this family is probably better represented in tropical waters than most of the other families.

Family Ammotheidae (Fig. 40b) *and Family Tanystylidae* (Fig. 40a)

These two families consist of a diverse group of species whose common character is the reduction of the chelifores in the adult. These are present in a chelate form in the immature forms but in the adults of most species they are achelate or in a few species where they

---

Fig. 40

The form of the body in the Tanystylidae, Ammotheidae, Phoxichilidiidae and in the fossils Palaeoisopus and Palaeopantopus.
(a) Tanystylidae.
(b) *Achelia echinata*.
(c) *Phoxichilidium femoratum*.
(d) *Anoplodactylus petiolatus*.
(e) *Palaeoisopus problematicus*. (After Broili, 1932)
(f) *Palaeopantopus*. (After Broili, 1932)

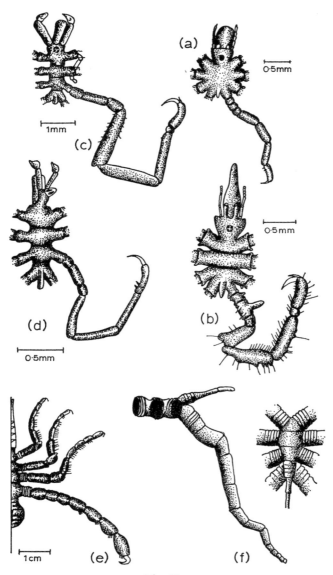

Fig. 40

remain chelate the chelae are small and lack teeth on the fingers. Both sexes have well-developed palpi and ovigers and the latter usually have compound spines on the terminal segments. Although some species are found at depths greater than 1000 metres, the majority are either littoral or shallow water forms. In many species, the proboscis is capable of considerable movement. The family Ammotheidae has doubtful validity and at least forty genera have been included within it at one time or another but of these, only about thirty are now considered as valid units. Suggestions have been made that some genera should be included in other families such as the Tanystylidae. The first of these has been considered here as a second family although they have many characters in common with the Ammotheidae. The only real differences are that the proboscis of the Ammotheidae is usually large and bulbous whilst that of the Tanystylidae varies in shape between species and may be quite attenuated. Also, the palps of the ammotheids have more segments than those of the Tanystylidae. A major revision of the relationships of the genera included arbitrarily in these families is needed. To complete the task satisfactorily, it would be necessary to assess something in the order of 200 species which would require a computer and, what is more important, accurate description of the species (Fry and Hedgpeth, 1969; Hedgpeth, 1947).

Speciation in Ammotheidae is complex and indistinct but within some areas, attempts to analyse the species present have been made using the techniques of numerical taxonomy. Most of the characters used depend upon the ratios of two dimensions, such as the length of the trunk compared with that of the proboscis. These are referred to as mensural characters and changes in their relationships to each other often occur when the pycnogonid reaches maturity. The ranges of variation in measurements within the species are so great that a complex pattern of overlap results which makes the separation of the species very difficult. When the morphological similarities of a number of species are expressed as a dendrogram the resulting picture is very complex. Obviously breeding experiments would be more likely to give a definitive result in a situation like this.

By some workers the genus *Austrodecus* belonging to the Tanystylidae is raised to the rank of family (Stock, 1957).

ORDER PALAEOPANTOPODA (1930) (Fig. 40e,f)

This order is represented by a single species, *Palaeopantopus maucheri*, which has been described from two specimens found in the Hunsrück shale from the Lower Devonian of western Germany (Fig. 40f) (Hedgpeth, 1955). The trunk has four segments and the lateral

Systematics

processes have a number of annular markings or swellings. The legs have three coxal segments, five longer segments and several short terminal segments but there is doubt as to whether or not gonopores are present on the legs. The abdomen, unlike that of living forms, consists of two or three segments. Near the anterior end of the animal there is a paired, segmented appendage, but whether this represents the ovigers or palps has not been determined. Chelifores are absent and there is no visible proboscis, but since only the dorsal surface is known, the possible existence of some sort of recurved, ventrally-borne proboscideal structure must be considered as a possibility. No significance can be placed on the absence of eyes and chelifores because these are absent in some living pycnogonids. The numerous small distal leg joints are reminiscent of the condition in the living genus *Nymphonella* (Ohshima, 1927a). The occurrence of two or three abdominal segments is interesting since this arrangement is comparable with the range of variation observed in the nervous system of some existing pycnogonids in which some have two pairs of abdominal ganglia, although there is no suggestion of externally visible segmentation. The segmented abdomen in the fossils has been cited as evidence for the gradual reduction of the abdomen from the most posterior segments during evolution of this group. A tendency for loss of abdominal segmentation has been observed in chelicerate groups and in these it was complete by the Devonian. If *Palaeopantopus* is accepted as a pycnogonid and the trend for reduction in the abdomen did in fact occur, one must then conclude that it was more or less complete before the mid-Devonian to correspond with the chelicerates. This would imply that instead of the pycnogonids being a relatively new group of arthropods, the living Pycnogonida may be survivors of one of the oldest arthropod groups.

Palaeoisopus problematicus (Fig. 40e), another fossil from the Hunsrück shale described from four specimens, was initially attributed, as its name suggest, to the Isopoda, but later because of an apparent resemblance to *Palaeopantopus maucheri* it was grouped with the Pycnogonida. The only similarities between the two fossils are the rings on the lateral processes and the generally attenuated body. An alternative suggestion is that *P. problematicus* is the larval form of some marine arachnid, but larval forms usually possess anterior appendages and the well-developed abdomen in this fossil suggests that it is a mature form. Also, it is unlikely that four similar larval specimens would be found without some associated remains of adults. The present view is that *P. problematicus* belongs amongst the arthropods but it is unlikely that it has a close relationship with the pycnogonids (Hedgpeth, 1955).

Subsequently, two investigators working independently, one using the analysis of morphological features (Dubinin, 1957) and the other a radiographic method (Lehman, 1959), concluded that the previous interpretation of the fossil material was incorrect. They suggested that the segmented anterior end is really the abdomen and the swollen portion represents a complex consisting of a type of proboscis, a pair of chelifores together with segmented appendages which are possibly palps. This is an interesting idea because if it is a fossil pycnogonid, then reduction of the abdomen had not progressed far at this time and the living forms may have evolved at a later date than was previously believed (Sharov, 1966).

Extra-legged species of the families Pycnogonidae, Nymphonidae and Colossendeidae

In contrast with the normal octopodous condition there are some species of pycnogonid which have five pairs of legs and one which has six pairs. Although the first species with more than eight legs was described in 1835, this was dismissed as an error or a monstrosity, until the South Polar expeditions at the turn of the century obtained more specimens. The cause and significance of this decapodous and dodecapodous condition is unknown. It occurs in three families with widely different shapes. The Nymphonidae, which have chelifores and palps; the Pycnogonidae, which are without either, and the Colossendeidae which have only palps. An interesting point concerning the extra-legged forms is that they occur in those families which have few genera but numerous species and not in families which have a relatively large number of genera compared with the species. This means that the extra-legged forms occur in the most homogenous families. It has been suggested that these extra-legged forms occur in those branches of the Pycnogonida which, because of the large numbers of numerically abundant species resembling each other closely, are the most successful from an evolutionary point of view. To date, they have only been found in the Antarctic and Caribbean, both of which are considered to have a rich pycnogonid fauna.

In each locality in which the extra-legged forms occur, there are also closely related octopodous forms. *Pentapycnon charcoti* occurs in the South Shetlands and *Pycnogonum gaini* in the Ross Sea. The last tubercle on the dorsal surface of *P. gaini* is a possible vestigial remnant of an extra segment. In these species the only differences between them, apart from the extra segment, is the shape of the proboscis. In the Caribbean, *Pentapycnon geayi* resembles an octopodous species of *Pycnogonum* (Fig. 38). *Pentanymphon antarcticum* (Fig. 39) differs from *Nymphon hiemale* by having a shorter tarsus, a longer third joint to the palp and possibly a difference in the shape of the spines of the ovigers. The similarity between the *Decolopoda* and *Colossendeis* is not so pronounced since the former retains the immature chelifores in the adult condition, although the latter usually sheds them. A closer similarity that this, however, exists between *Colossendeis wilsoni* and *Decalopoda antarcticum* in which the former differs from the other members of this genus by having eight segments in its palp instead of the usual number, nine. The species with extra legs frequently have a wider geographical and bathymetric range than the octopodous forms which may suggest that they are more successful. An exception to the usual pattern is *Pentacolossendeis reticulata,* which occurs at about the 100 fathom line, south of Florida Keys, without a closely related *Colossendeis* in this locality.

In general the extra-legged species are more numerous south of the region known as the Antarctic–south Atlantic convergence. When the existence of the extra-legged forms had been accepted, the problem of their phylogenetic significance was considered. Two main views were expressed: one considered that these forms represent the original, primitive state of the pycnogonids and the other that they are a phenomenon arising from the octopodous form. It has been suggested that since five ganglia are present in the octopodous forms, one of these may relate to a lost segment. It is possible that under some unspecified conditions this segment is not always lost. A single larva of *Nymphon spinosum* from the north Pacific was collected which had a fifth pair of rudimentary legs but this may have been the result of faulty development since there was no suggestion of the development of a fifth trunk segment. Although regions of metameric instability during development in both the anterior and posterior parts of the body have been postulated, no conclusion has yet been reached on this problem.

It is possible, because of the overlap in their areas, that the extra-legged specimens are variants of the octopodous forms, together forming a polymorphic species, but nothing is known at present concerning the genetics of these species so this possibility has not been investigated.

KEY TO THE ADULTS OF PYCNOGONID FAMILIES

A. *Four pairs of ambulatory appendages*

1. Chelifores and palpi present   2
   Chelifores or palpi, or both, lacking   3

2. Palpi 5-jointed, large chelae which over-reach the proboscis   NYMPHONIDAE

   Palpi 6/10-jointed, chelae small, chelifores shorter than the proboscis which is usually large and bulbous   4

   Pairs of ambulatory appendages   AMMOTHEIDAE

   Palps 4/6-jointed, chelifores shorter than the proboscis, proboscis more elongate   TANYSTYLIDAE

3. Chelifores present, palpi lacking   4
   oth chelifores and palpi lacking   5
   Palpi present, chelifores lacking   COLOSSENDEIDAE

4. Ovigers 10-jointed, in both sexes   PALLENIDAE
   Ovigers 5/6-jointed, in male only   PHOXICHILIDIIDAE

5. Body slender, legs about twice as long as the body. Auxiliary claws present. Ovigers 7-jointed in the male only   ENDEIDAE

   Body stout, legs short, little longer than the body. Auxiliary claws absent. Ovigers 9-jointed, in male only   PYCNOGONIDAE

B. *Five or six pairs of ambulatory appendages*

1. Five pairs of walking legs   2
   Six pairs of walking legs   DODECOLOPODA

2. Without palps or chelifores   PENTAPYCNON
   With palps and with chelifores   3

3. Body slender and elongate   PENTANYMPHON
   Body disc shaped   4

4. Lateral processes widely separated at their bases: intersegmental sutures present on the trunk; no chelifores in adults   PENTACOLOSSENDEIS

   Lateral processes contiguous over at least the basal half of their lengths; no intersegmental sutures on the trunk; chelifores well developed and chelate in adults   DECOLOPODA

BIBLIOGRAPHY

Allman, G. J., 1862. 'On a remarkable form of parasitism among the Pycnogonidae', *Ann. Mag. nat. Hist.*, **9**(3), 36–43.
Arita, K., 1937. 'Beitrage zur biologie der pantopoden', *J. Dep. Agric. Kyushu imp. Univ.*, **5**(6), 271–88.
Baccetti, B., and Rosati, F., 1971. 'Electron microscopy on Tardigrades. III Integument', *J. Ultrastruct. Res.* **34**, 214–43.
Barnard, K. H., 1954. 'South African Pycnogonida', *Ann S. Afr. Mus.*, **41**(3), 81–158.
Beams, H. W., and Kessel, R. G., 1963. 'Electron microscope studies on developing crayfish oocytes with special reference to the origin of yolk', *J. Cell Biol.*, **18**, 621–49.
Benson, P. H., and Chivers, D. C., 1960. 'A pycnogonid infestation of *Mytilus californianus*', *Veliger*, **3**(1), 16–18.
Bouvier, E. L., 1923. 'Pycnogonides', *Faune de France*, **7**, 1–69.
Broili, F., 1932. 'Palaeoisopus ist ein Pantopode', *Sitzungsb. Bayer. Akad. Wiss. (Math-Naturh. abt)*, 45–60.
Bullock, T. H., and Horridge, G. A., 1965. *Structure and function in the nervous systems of invertebrates*, W. H. Freeman and Company, London, 1–1611.
Calman, W. T., 1923. 'Pycnogonida of the Indian Museum', *Rec. Indian Museum*, **25**(3), 265.
Carpenter, G. H., 1904. 'Pycnogonida', *Fish., Ireland, Sci. Invest.*, **4**, 1905, 1.
Clark, W. C., 1956. 'A new species of *Pycnogonum* from Banks Peninsula, New Zealand', *Rec. Canterbury Mus.*, **7**(2), 171–3.
Clark, W. C., 1963. 'Australian Pycnogonida', *Records of the Australian Museum.*, **26**(1), 1–81.
Cole, L. J., 1901. 'Notes on the habits of pycnogonids', *Biol. Bull. Mar. biol. Lab.*, Woods Hole, **2**(5), 195–207.
Cole, L. J., 1904. 'Pycnogonida collected at Bermuda', *Proc. Boston Soc. nat. Hist.*, **31**, 315–28.
Cole, L. J., 1906. 'Feeding habits of the pycnogonid *Anoplodactylus lentus*', *Zool. Anz.*, **26**, 740–1.

Cole, L. J., 1910. 'Peculiar habitat of a pycnogonid (*Endeis spinosus*) new to North America, with observations on the heart and circulation', *Biol. Bull. Mar. biol. Lab.*, Woods Hole, **18**(4), 193–203.

Crapp, G. B., 1968. 'The identification and ecology of Anglesey pycnogonids', M.Sc. thesis: Bangor University.

Dawydoff, C., 1928. *Traite embryologie comparée des invertebrates*, Masson and Cie., France, 1–930.

Dearborn, J. H., 1967. 'Food and reproduction of *Glyptonotus antarcticus* (Crustacea: Isopoda) at McMurdo Sound, Antarctica', *Trans. Roy. Soc. New Zealand*, Zoology **8**(15), 163–8.

Dogiel, V., 1911. 'A short account of work on Pycnogonida done during June, 1911, at Cullercoats, Northumberland Sea Fish. Comm.', *Rep. Sci. Invest.*, 26–7.

Dohrn, A., 1881. 'Die Pantopoden des Golfes von Neapel und der angrenzenden Meersabschnitte', *Fauna G. Flora nolf Neapel*, Vol. 3.

Droller, M. J., and Roth, T. F., 1966. 'An electron microscope study of yolk formation during oogenesis in *Lebistes reticulatus guppyi*', *J. Cell. Biol.*, **28**, 209–33.

Dubinin, V. B., 1957. 'On the orientation of the cephalic end of the Devonian pycnogonids of the genus *Palaeoisopus* and their systematic position in the Arthropoda', *Doklady Akademiy Nauk U.S.S.R. Tom.*, **117**(5), 881–4.

Dumont, J. N., and Anderson, E., 1967. 'Vitellogenesis in the horseshoe crab *Limulus polyphemus*', *J. microscopie*, **6**, 791–806.

Fage, L., 1932. 'Peches planctoniques à la lumière effectueés à Banyuls-sur-Mer et à Concarneau', *Archives de Zoologie Experimentale et Generale*, **74**, 249–61.

Fage, L., 1949. 'Classe de Pycnogonides', *Traité de Zoologie*, **6**, 906–41.

Fage, L., 1954. 'Remarques sur les pychogonides abyssaux', *Un. int. Sci. biol. Ser. B.*, **16**, 49–56.

Flym, T. T., 1928. 'The Pycnogonida of the marine survey of South Africa', *Fish. mar. Biol. Surv.*, rep 6, spec. rep 1.

Friedrich, H., 1969. *Marine Biology*, Sidgwick and Jackson, London, 1–474.

Fry, W. G., 1965. 'The feeding mechanisms and preferred foods of three species of Pycnogonida', *Bull. Br. Mus. Nat. Hist.*, D.**12**(6), 197–223.

Fry, W. G., and Hedgpeth, J. W., 1969. 'The Fauna of the Ross Sea, Part 7, Pycnogonida, 1 Colossendeidae, Pycnogonidae, Endeidae, Ammotheidae', *New Zealand Oceanographic Institute Memoir*, No. 49, 1–139.

Giltay, L., 1928. 'Note sur les pycnogonides de la Belgique', *Bull. Annls. Soc. r. ent. Belg.*, **68**, 193–229.

Giltay, L., 1929. 'Quelques pycnogonides des environs de Banyuls (France)', *Bull. Annls. Soc. r. ent. Belg.*, **69**, 172–6.

Gordon, I., 1932. 'Pycnogonida', *Discovery Rep.*, **6**, 1–138.

Graber, V., 1880. 'Ueber das unicorniale Tracheaten-und speciell das Arachnoideen-und Myriopoden-Auge', *Arch. f. Mikr. Anatomie*, **XVII**, 58–93.

Grenacher, H., 1879. *Untersuchungen uber das Sehorgan der Arthropoden*, **4**, Gottingen.

Hanstrom, B., 1965. 'Indications of neurosecretion and the structure of Sokolow's organ in pycnogonids', *Sarsia*, **18**, 24–36.

Hedgpeth, J. W., 1941. 'A key to the Pycnogonida of the Pacific Coast of North America', *Trans. S. Diego Soc. Nat. Hist.*, **9**(26), 253–64.

Hedgpeth, J. W., 1947. 'On the evolutionary significance of the Pycnogonida', *Smithson misc. Coll.*, **106** (18).

Bibliography

Hedgpeth, J. W., 1949. 'Report on the Pycnogonida collected by the albatross in Japanese waters', *Proc. U.S. nat. Mus.*, **98**, 233–321.

Hedgpeth, J. W., 1951. 'Pycnogonids from Dillon Beach and vicinity, California, with descriptions of two new species', *Wasmann. J. Biol.*, **9**(1), 105–17.

Hedgpeth, J. W., 1955. 'Pycnogonida', *Treatise on Invertebrate Palaeontology*, R. C. Moore, Ed. Geological Society of America and University of Kansas Press, Part P *Arthropoda* **2**, 163–73.

Hedgpeth, J. W., 1963. 'Pycnogonids of the North American Arctic', *J. Fish. Res. Bd. Can.*, **20**(50), 1315–48.

Hedgpeth, J. W. 1964. 'Notes on the peculiar egg-laying habit of an Antarctic prosobranch (Mollusca: Gastropoda)', *Veliger*, **7**(1), 45–6.

Helfer, H., and Schlottke, E., 1935. 'Pantopoda', *Bronn's Kl. Ordn. Tierreichs*, **5**, IV(2).

Hilton, W. A., 1934. 'Notes on parasitic pycnogonids', *J. Ent. Zool.* (Pomona Coll.), **26**(4), 57.

Hoek, P. P. C., 1881. 'Report on the Pycnogonida dredged by H.M.S. *Challenger* during the years 1893–96', *Rep. Scient. Voy. H.M.S. Challenger*, **3**, 1–252.

Isaac, M. J., and Jarvis, J. H., 'Endogenous tidal rhythmicity in the littoral pycnogonid *Nymphon gracile* (leach)', *J. exp. mar. Biol. & Ecol.* (in press).

Jarvis, J. H., 1972. 'Aspects of the reproductive biology of pycnogonids', Ph.D. thesis, Swansea.

Jarvis, J. H., and Isaac, M. J. 'Bioluminescence in *Nymphon gracile*' (personal communication).

Jarvis, J. H., and King, P. E., 1972a. 'Reproduction and development in the pycnogonid *Pycnogonium littorale*', *Mar. Biol.*, **13**, 146–55.

Jarvis, J. H., and King, P. E., 1972b. 'Ultrastructure of the eye of *Nymphon gracile* (in preparation).

Kessel, R. G., 1968. 'Annulate Lamellae,' *J. Ultrastruct. Res. Supplement*, **10**, 5–82.

King, P. E., Bailey, J. H., and Babbage, P. C., 1969. 'Vitellogenesis and formation of the egg chain in *Spirorbis borealis* (Serpulidae)', *J. mar. biol. Ass. U.K.*,' **49**, 141–50.

King, P. E., and Crapp, G. B., 1971. 'Littoral pycnogonids of the British Isles', *Field Studies*, **3**, 455–8.

King, P. E., and Jarvis J. H., 1970. 'Egg development in a littoral pycnogonid *Nymphon gracile*', *Mar. Biol.*, **7**, no. 4, 294–304.

Knight-Jones, E. W., and Macfadyen, A., 1959. 'The metachronism of limb and body movements in annelids and arthropods', *Proc. XVth Int. Congr. Zool.*, 969–71.

Krishnan, G., 1955. 'Nature of the cuticle of Pycnogonida', *Nature*, **175**, 904.

Lebour, M. V., 1916. 'Notes on the life-history of *Anaphia petiolata* (Kröyer)', *J. mar. biol. Ass. U.K.*, **11**, 51–6.

Lebour, M. V., 1947. 'Notes on the Pycnogonida of Plymouth', *J. mar. biol. Ass. U.K.*, **26**, 139–65.

Lehman, W. M., 1959. 'Nene Entdeckungen an Palaeoisopus', *Paläont.* **2**, 33, 96–103.

Lochhead, J. H., 1961. 'Locomotion', in *Physiology of Crustacea*, Vol. 2 (Ed. T. H. Waterman), 313–64. New York and London: Academic Press.

Loman, J. C. C., 1907. 'Biologische Beobachtungen an einen Pantopoden', *Tijdschr. ned. dierk. Vereen* **2**(10), 255–84.

Lotz, Von Guntram, and Bückmann, D., 1968. 'Die Häutung und die Exuvie von *Pycnogonum litorale* (Ström) (Pantopoda)', *Zool. Jb. Anat. Bd.*, **85**, 529–36.

Marcus, E. du B. R., 1952. 'A hermaphrodite pantopod', *Anais. Acad. bras. Cienc.*, **24**(1), 23–30.

Morgan, E., 1971. 'The swimming of *Nymphon gracile* (Pycnogonida)', *J. exp. Biol.*, **55**, 273–87.

Morgan, E., Nelson-Smith, A., and Knight-Jones, E. W., 1964. 'Responses of *Nymphon gracile* (Pycnogonida) to pressure cycles of tidal frequency', *J. Exp. Biol.* **41**, 825–36.

Morgan, Th. H., 1891. *A contribution to the embryology and phylogeny of the Pycnogonids*, Stud. Biol. Labor. Johns Hopkins Univ., **5**.

Nørrevang, A., 1965. 'Oogenesis in *Priapulus candatus* Lamarck', *Meddr. dansk naturh. Foren.*, **128**, 2–76.

Nørrevang, A., 1968. 'Electron microscopic morphology of oogenesis', *Int. Rev. Cytol.*, **23**, 114–76.

Ohshima, A., 1927a. '*Nymphonella tapetis*, n.g., n.s.p., a pycnogon parasitic in a bivalve', *Annotnes. zool. Jap.*, **11**(3), 257–63.

Ohshima, H., 1927b. 'Notes on some pycnogons living semi-parasitic on holothurians', *Proc. imp. Acad. Japan*, **3**(9), 610–13.

Ohshima, H., 1933. 'Young pycnogonids found parasitic on nudibranchs', *Annotnes. zool. Jap.*, **14**(1), 61–6.

Okuda, S., 1940. 'Metamorphosis of a pycnogonid parasitic in a hydromedusa', *J. Fac. Sci., Hokkaids Imp. Univ.* (6) *Zool.*, **7**(2), 73–86.

Prell, H., 1910. 'Beiträge zur Kenntniss der Lebensweise einiger Pantopoden', *Bergens Mus. Aarbok*, **10**.

Raven, C. P., 1961. *Oogenesis, the storage of developmental information*, Pergamon Press, London.

Ricketts, E. F., and Calvin, J., 1960. *Between Pacific Tides*, Stanford U.P., Stanford, 3rd. ed.

Sanchez, S., 1959. 'Le developpement des Pycnogonides et leurs affinities avec les Arachnides', *Archs. Zool. exp. gen.*, **98**(1), 1–101.

Schlottke, E., 1933. 'Darm und Verdauung bei Pantopoden', *Zs. f. mikroskopanat. Forschung*, **32**(4), 633–58.

Sharov, A. G., 1966. *Basic Arthropoden Stock with Special Reference to Insects*, London: Pergamon Press, 1–271.

Slifer, E. H., 1969, 'Sense organs on the antenna of a parasitic wasp *Nasonia vitripennis* (Hymenoptera, Pteromalidae)', *Biol. Bull.*, **136**, 253–63.

Sokolow, I., 1911. 'Uber den Bau der Panterpodenanger', *Z. wiss. Zool.*, **98**.

Stephensen, K., 1933. 'Pycnogonida. The Godthaab Expedition, 1928', *Medd. om. Groenland*, **79**(6).

Stephensen, K., 1937. 'Pycnogonida', *Zoology Iceland*, **3**(58).

Stock, J. H., 1954. 'Pycnogonida from the Indo-West-Pacific, Australia and New Zealand waters', *Vidensk. Medd. fra. Dansk naturh. Foren. Bd.*, **116**. 1–168.

Stock, J. H., 1955. 'Pycnogonida from the West Indies, Central America and the Pacific coast of North America', *Vidensk. Medd. fra. Dansk naturh. Foren. Bd.*, **117**, 209–66.

Stock, J. H., 1957. 'The pycnogonid family Austrodecidae', *Beaufortia*, **6**, no. 68, 1–81.

Stock, J. H., 1959. 'On some South African pycnogonids of the University of Cape Town Ecological Survey', *Trans. Roy. Soc. S. Africa*, **35**(5), 549–67.

Bibliography

Stock, J. H., 1963. 'Israel South Red Sea Expedition 1962, Report no. 3: Pycnogonida', *Bull. Sea Fish. Res. Sta. Haifa,* **35**, 27–34.

Storch, V., and Welsch, U., 1972. 'The ultrastructure of epidermal cells in marine invertebrates (Nemertini, Polychaeta, Prosobranchia, Opisthobranchia)', *Mar. Biol.,* **13**(2), 167–76.

Thompson, D'Arcy W., 1909. 'Pycnogonida', in *The Cambridge Natural History,* vol. 4, 501–42. Ed. by S. F. Harmer and A. E. Shipley, London: Macmillan & Co.

Utinomi, H., 1954. 'The fauna of Akkeshi Bay, XIX. Littoral Pycnogonida', *Publ. Akkeshi Mar. Biol. Stat.,* no. **3**, 1–28.

White, R. H., 1967. 'The effect of light and light deprivation upon the ultrastructure of the larval mosquito eye. II The Rhabdom', *J. exp. Zool.,* **166**, 405–26.

Wilson, E. B., 1880. 'The Pycnogonida of New England and adjacent waters', *Rep. U.S. Comm. Fisheries,* pt. 6., 463–504.

Wiren, E., 1918. 'Zur morphologie und Phylogenie der Pantopoden', *Zool. Bidrag. fran Uppsala,* **6**, 41–181.

Wyer, D., 1972. 'Studies on the nutritional biology of pycnogonids', Ph.D. thesis, Swansea.

Wyer, D., and King, P. E. 'Studies on feeding in British pycnogonids'. (in preparation).

Zeigler, A. C., 1960. 'Annotated list of Pycnogonida collected near Bolnias, California', *Veliger,* **3**(1), 19–22.

INDEX OF SUBJECTS

abdomen, 7, 9, 10, 46, 49, 113, 129
Alaska, 109
alimentary canal, 16, 22, 46, 49, 78, 80, 123
anal segment (*see* abdomen)
anemone, 37, 91
Antarctic, 8, 99, 106, 109, 130
anus, 7, 10, 17, 46, 49
Arctic, 13, 99, 101
arthrodial membrane, 14, 23
Atlantic, 99, 101, 103, 110, 124
Australia, 106, 109
auxiliary claw, 10
Azores, 123

Bering Sea, 99, 101
Bermuda, 108
brain, 53
Bristol Channel, 35
British Isles, 124
Black Sea, 123
blastopore, 75
burrowing, 38

caeca, 39, 46, 49, 51, 64, 88, 113
California, 14, 91, 108
Caribbean, 108, 120, 130
cells, 44, 48, 54
cement, 28, 121
cement gland, 43, 45, 116
cephalon, 9, 21, 24
cephalothorax, 9, 50
chela, 24, 25, 89
chelicera, 24

chelifores, 7, 9, 10, 13, 24, 25, 26, 32, 45, 77, 80, 85, 129
circulation, 49, 50
circumoesophageal commisure, 21, 22, 53
claw, 10
cleavage, 75
coelom, 52, 77
colour, 39
Concarneau, 35
connective tissue, 49
corpuscles (*see* haemocytes)
coxa, 10, 31, 66
cuticle, 9, 11, 14, 16, 41, 43, 51, 59, 90, 114, 123

Davis Strait, 101
decapodous species, 110, 130
deutocerebrum, 54, 113
development, 56, 82
digestion, 8, 48
distribution, 26, 35, 38, 91, 99, 101, 103, 106, 108, 123, 124, 130
diverticulum (*see* caecum)
dorsal vessel, 51, 80
Drake Passage, 104

East Indies, 108
ecdysis, 16
ectoderm, 77, 83
egg, 28, 35, 45, 66, 74, 80, 90, 121
embryo, 80
endocrine system (*see* hormone)
endoderm, 75, 80

Index of subjects

endogenous rhythm, 35
enzymes, 88
epiphytes, 84
epizooites, 37, 84
evolution, 112
excretion, 8, 44, 49, 52, 63
eye, 56, 113
eye tubercle, 9, 10, 17, 22, 50, 56, 113, 120

Falkland Islands, 104
fecundation, 72
feeding, 8, 11, 13, 16, 20, 84, 123
female, 26, 29, 116
femur, 28, 72
fertilisation, 72
Florida, 124
foregut, 14, 16, 22
fossil, 128

Galapagos, 14
gall, 90, 93
ganglia, 21, 77, 83, 113
gastrulation, 75
genital opening, 31, 32, 65, 71, 72, 113, 121
germ layer, 75
gland, 44, 45, 48
gonad, 17, 39
granulocyte, 57
Greenland, 101
grooming, 88
gut epithelium, 48

haemocoele, 50
haemocyte, 80
haemolymph, 50, 52, 68
hair (see seta)
hermaphrodite, 64, 65
hindgut, 46, 49
hormone, 52, 54
hypodermis, 44, 46, 50, 60

India, 108
integument, 41

Japan, 26, 108, 109
jaw, 11
juvenile, 12, 31, 99, 120, 123

Korea, 108

larva, 24, 26, 28, 45, 52, 54, 58, 73, 80, 90, 126

lateral projection, 9, 10
legs, 9, 14, 31
leucocyte, 51
life cycle, 72
locomotion, 11, 29, 31, 38, 99, 106

Malaya, 108, 109
male, 26, 29
mating, 72
Mediterranean, 26, 108, 110
mesoderm, 77
midgut, 16, 17, 46, 49, 80, 88
migration, 104
moulting, 39, 43, 44, 82
mouth, 7, 13, 14, 16, 17, 24, 46, 86
mucus, 28, 46, 114
muscles, 14, 16, 18, 19, 20, 21, 22, 31, 48, 49, 53, 71, 88

Naples, 124
nauplius, 80
nervous sytem, 17, 18, 21, 22, 53, 77
neurosecretion, 54
New Zealand, 106
North America, 109
Norway, 123, 124

oesophagus, 46
Okhotsk Sea, 101
oocyte (see egg)
oogenesis, 65, 112
operculum, 66
opisthosoma, 7
ova (see egg)
ovary, 51, 64, 65
ovigeral spine, 25
ovigerous legs, 7, 8, 9, 10, 25, 26, 28, 31, 32, 62, 72, 80, 88, 113, 116, 129
ovigers (see ovigerous legs)

Pacific Ocean, 101, 110, 131
palps, 7, 9, 10, 26, 31, 32, 39, 45, 64, 80, 129
Panama, 109
pharynx, 11, 14, 16, 17
phosphorescence, 39
plankton, 35, 37, 124
plummeting, 34
Point Barrow, 99
pressure, 35, 52, 62
proboscis, 7, 9, 10, 11, 13, 15, 16, 17, 19, 20, 21, 22, 24, 26, 32, 46, 53, 78, 85, 97, 114, 128
proctodeum, 52, 80

propodus, 10, 28, 49
prosoma, 7
protocerebrum, 54, 77

regeneration, 38, 43
reproduction, 34, 64, 120
respiration, 8, 44, 50, 52
retina, 6
Ross Sea, 104
Russia, 125

Sagami Bay, 99
Sargassum, 38, 108, 110, 123
scape, 24
segmentation, 32, 113, 114, 129
senses: tactile, 13, 26; chemoreception, 13, 62, 64; visual, 56, 62
seta, 9, 11, 16, 26, 62, 63, 89
sex, 26, 64
Singapore, 109
Sokolow's organ, 54, 63
South Africa, 108
South America, 106, 123
South Georgia, 104
sperm, 65
stomodeum, 78
Suez Canal, 115

supra-oesophageal ganglion, 53
swimming, 31, 33
symbionts, 39
systematics (*see* taxonomy)

tapetum, 61
tarsus, 10, 28
taxonomy, 99, 115
testis, 64
tibia, 10
Torres Strait, 109
trunk, 9, 10, 32

ultrastructure, 43, 59, 61, 63, 66, 67

valvula, 46, 49
ventral organ, 56
vitellogenesis, 66, 67, 112
vitellophages, 77

walking, 31
West Indies, 108, 110

yolk, 65, 75, 80, 90

INDEX OF SPECIES

Abietinaria, 96
Achelia, 26, 28, 52, 64, 99, 104, 106, 107, 117
A. alaskensis, 92
A. assimilis, 106
A. australiensis, 106
A. besnardi, 106
A. borealis, 101
A. chelata, 94
A. communis, 106
A. dohrni, 106
A. echinata, 9, 11, 14, 16, 26, 29, 34, 42, 49, 62, 63, 66, 74, 81, 86, 89, 94, 97, 99, 111, 126
A. fernandeziana, 106
A. gracilis, 94, 106
A. hispida, 26
A. hoekii, 106
A. intermedia, 99
A. laevis, 9
A. latifrons, 108
A. lavrentii, 99
A. litke, 99
A. longipes, 11, 26, 42, 84, 89, 98
A. nudiuscula, 94
A. parvula, 106
A. sawayai, 106
A. serratipalpis, 106
A. simplex, 26, 64
A. spicata, 106
A. spinosa, 99, 111
A. sufflata, 106
A. uschakovi, 99
A. variabilis, 56, 57

Actinia equina, 86, 96, 97
Aglaophenia, 39, 92, 94, 95, 96
Ainigma, 117
Alcyonium, 96
Algae, 89
Ammothea, 24, 45, 53, 75, 92, 104
A. clausi, 106
A. discata, 94
A. glacialis, 12, 97
A. longispina, 12, 13
A. magniceps, 106
A. minor, 106
Ammotheidae, 10, 11, 13, 24, 32, 88, 89, 90, 99, 115, 117, 128, 132
Ammothella appendiculata, 43, 56, 57
A. biunguiculata, 108, 111
A. heterosetosa, 110
A. regulosa, 110
Anammothea, 104
Anodonta, 69
Anoplodactylus, 28, 31, 75, 82, 110
A. angulatus, 24, 25, 39, 84, 86, 95
A. erectus, 93, 95
A. exiguus, 90, 93
A. grandulifer, 108
A. lentus, 34, 39, 72, 86, 95, 97
A. pelagicus, 38
A. petiolatus, 38, 88, 90, 93, 95, 126
A. portus, 110
A. pygmaeus, 93, 95
A. saxatiles, 108
A. typhlops, 38, 43
Anoropallene valida, 109

Index of species

Anthopleura xanthogrammica, 86, 96, 97
Antennularia, 89, 98, 123
Arachnids, 58, 112
Arbacia, 68
Armina varidosa, 92
Artemia, 68
Arthropoda, 112
Ascophyllum, 39, 95
Ascorhynchus, 24, 42, 45, 104, 108
A. agassizi, 110
A. arenicola, 45
A. armatum, 108, 110
A. auchenicum, 11, 56, 57
A. castelli, 45
A. corderoi, 64, 98
A. melwardi, 109
A. minutus, 109
Astacus, 69
Audoinia, 95
Austrodecidae, 117
Austrodecus, 128
A. breviceps, 56, 57
A. frigorifugum, 56, 57
A. glaciale, 13, 18, 20, 22, 23, 86, 89, 96
A. simulans, 56, 57

Bohmia chelata, 94
Boreonymphon robustum, 24, 25, 45
Bougainvillea, 86
Bowerbankia, 86, 95, 97
Bunodactis elegantissima, 86, 96, 97

Callipallene, 24, 26, 66, 72, 75, 76, 113
C. brevirostris, 11, 34, 39, 62, 124
C. californiensis, 110
C. emaciata, 110
C. phantoma, 124
Campanularia, 86, 97
Cephalocarida, 112
Cellarinella, 86, 96, 97
C. foveata, 86
Chaetonymphon, 54, 75
C. macronyx, 25
Cheilopallene, 110
Cirripedia, 112
Clava, 86, 93, 96
Clavelina, 96
Clotenia conirostris, 9
Colossendeidae, 32, 90, 117, 130, 132

Colossendeis, 24, 26, 28, 38, 49, 62, 101, 104, 106, 110, 115, 117, 118, 120, 131
C. australis, 105, 118
C. colossea, 39
C. drakei, 12, 105
C. frigida, 25
C. lilliei, 116
C. longirostris, 105, 116, 118
C. megalonyx, 105, 116, 118, 120
C. orcadense, 120
C. proboscidea, 64, 94, 100, 101
C. robusta, 105
C. scotti, 105
C. stramenti, 118, 120
C. tortipalpis, 105, 116, 118, 120
C. wilsoni, 105, 131
Corallinae, 39
Cordylochele, 24
C. longicollis, 24, 25
Corymorpha, 95
Coryne, 86, 92, 93
C. muscoides, 90
Cosmetira, 93
Crustacea, 58, 80, 112, 113
Cucumaria, 97
C. frondosa, 86

Decachela, 109, 111, 117
D. discata, 108, 109, 122
Decalopoda, 117, 118, 120, 131
D. antarcticum, 131
D. australis, 116
Dodecalopoda, 117, 118, 120, 132, 133
Dynamena pumila, 86, 88, 89, 97

Ecleipsothremma spinosa, 106
Endeidae, 13, 24, 32, 89, 90, 115, 117, 121, 132
Endeis, 24, 26, 28, 38, 50, 123
E. flaccida, 108
E. laevis, 13, 72, 89
E. meridionalis, 108
E. mollis, 37
E. spinosa (= spinosus), 11, 13, 14, 16, 17, 37, 38, 46, 48, 49, 54, 62, 63, 72, 79, 84, 86, 92, 94, 98, 122, 123
Ephyrogymna, 110
Eudendrium, 39, 86, 94, 95, 96, 97
E. ramosum, 86
Eunephthys, 94
Eurycide, 24, 26, 42
E. longisetosa, 110
E. raphiaster, 110

Index of species

Flustra foliacea, 11, 16, 26, 84, 86, 89, 94, 97
Flustrellidra, 84

Gigartina, 84
Glyptonotus antarcticus, 91
Gorgonia, 94

Halichondria, 95
Halosoma viridintestinale, 95
Hannonia, 95
Hemichela, 108
Heterofragila, 108
Holothuria lubrica, 92, 94
Hydroids, 94

Insects, 114

Laomedea angulata, 84, 88
Lebistes, 69
Lectythorhynchus, 109, 111
L. hilgendorfi, 92
L. marginatus, 92, 94, 108, 109

Metridium dianthus, 86, 97
M. senile, 86, 96, 97
Milne-Edwardsia loweni, 97
Myriapoda, 82
Mytilus, 69, 94

Neonymphon, 110
Nucella, 98
Nymphon, 24, 31, 34, 45, 66, 70, 80, 108, 115, 117
N. bipunctatum, 38
N. brevicaudatum, 80, 92
N. brevirostre, 13, 25, 125
N. gracile, 9, 11, 13, 25, 26, 29, 31, 34, 35, 37, 40, 42, 43, 44, 48, 49, 50, 60, 61, 62, 64, 66, 68, 69, 74, 75, 86, 88, 125
N. grossipes, 101, 109
N. hamatum, 45
N. hiemale, 130
N. hirtipes, 92, 94, 100, 101
N. hirtum, 44, 91
N. leptocheles, 25, 38, 86, 98, 101
N. longitarse, 100, 101, 109
N. maculatum, 94
N. mixtum, 34
N. natalense, 38
N. parasiticum, 93, 95
N. pixellae, 53, 54
N. robustum, 92, 95, 101

N. rubrum, 13, 49, 50, 84, 88, 95, 125, 126
N. sluiteri, 97, 101
N. spinosum, 131
N. stylops, 58
Nymphonella, 108, 109, 111, 129
N. tapetis, 25, 26, 29, 38, 92
Nymphonidae, 10, 13, 32, 88, 89, 91, 115, 117, 123, 130, 132
Nymphopsis mucosa, 108

Obelia, 92, 94, 95, 123
O. dichotoma, 38, 39, 123
O. marginata, 39
Oorhynchus aucklandiae, 109
Oropallene, 109

Palaeoisopus problematicus, 126, 129
Palaeopantopoda, 115
Palaeopantopus, 126
P. maucheri, 128, 129
Pallene, 31, 37, 45, 76, 77, 80, 92
P. acus, 101
P. empusa, 95
Pallenidae, 26, 32, 91, 109, 115, 116, 117, 132
Pallenopsis, 26, 108, 115
P. calcanea, 101
P. hoeki, 109
P. molissima, 108
P. ovalis, 108
P. sibogae, 108
P. temperans, 108
P. tydemanni, 108
P. virgatus, 108
Pantopipetta, 117, 118
Paphia philipparum, 92
Paranymphon, 109, 111, 117, 125, 133
P. spinosum, 122
Parapallene bermudensis, 108
P. capillata, 11, 57
Patella coerulea, 69
Pentacolossendeis, 110, 117, 120
P. reticulata, 131
Pentanymphon, 117, 130
P. antarcticum, 25, 105
Pentaphynogonum, 117
Peripatus, 56
Phalangidae, 113
Phialidium hemisphericum, 38, 93
Phoxichilidiidae, 32, 88, 89, 115, 116, 117, 132
Phoxichilidium, 24, 26, 39, 45, 50, 64, 66, 72, 75, 82, 84, 101, 117

P. femoratum, 48, 49, 54, 63, 64, 126
P. hokkaidoense, 95
P. tubulariae, 93, 95, 123
P. virescens, 39, 93, 95
Phoxichilus, 123
P. charybdaeus, 51
Picrogromitus, 115
P. timosanus, 122
Pipetta weberi, 116, 118
Planorbis, 69
Podocoryne, 93
Polychaetes, 86, 112, 113
Polyorchis karatutoensis, 92
Porifera, 94
Priapulus, 68, 69
Protura, 82
Pseudopallene, 38, 115
P. circularis, 101
P. hospitalis, 108
Pycnogonidae, 32, 90, 115, 117, 121, 123, 130, 132
Pycnogonum, 24, 26, 75, 82, 101, 104, 110, 121, 123, 130
P. anovigerum, 121
P. benokianum, 95
P. gaini, 121, 130
P. hancocki, 14, 16, 17, 62, 90
P. littorale, 11, 14, 26, 44, 45, 49, 64, 66, 72, 73, 81, 84, 86, 90, 93, 96, 97, 122
P. megallanicum, 121
P. planum, 11, 56, 57
P. platylophum, 121
P. portus, 121
P. reticulatum, 110
P. rhinoceros, 11, 121
P. rickettsi, 14, 96, 97
P. stearnsi, 13, 18, 22, 23, 86, 91, 96, 97, 108

Rhopalorhynchus, 118
R. kröyeri, 109, 116, 118

Rhynchothorax, 24, 28, 115
R. australis, 13, 14, 16, 18, 21, 22, 23, 96, 97
R. kröyeri, 108
R. mediterraneus, 122
R. philopsammum, 38

Scipiolus, 108
Scolopendra, 56
Sertularia, 95
Sexanymphon, 117
Spider, 69
Spinther, 113
Spirorbis borealis, 68, 69
Sponge, 97
Stomotoca, 93
Syncoryne, 86, 93, 97
S. exima, 90

Tanystylidae, 32, 90, 115, 117, 128, 132
Tanystylum, 24, 45, 53, 66, 108, 110
T. anthomasti, 96, 101
T. calicirostre, 110
T. californicum, 39, 96
T. duospinum, 110
T. excuratum, 56, 57
T. gemium, 110
T. intermedium, 39, 96
T. isthmiacum, 110
T. orbiculare, 38, 39, 110
Tardigrada, 113
Tealia crassicornis, 86, 96, 97
Tethys leporina, 93, 95
Trygaeus, 24
Tubularia larynx, 86, 93, 95, 97, 116, 123
Turris, 93

Umbellula, 95

KING, PHILIP ERNEST
PYCNOGONIDS

000236159

HCL   QL447.K51

WITHDRAWN FROM STOCK
The University of Liverpool